Measurement of Safety Performance

Dan Petersen

American Society of Safety Engineers ❖❖ Des Plaines, Illinois

Published by the Ameican Society of Safety Engineers, Des Plaines, Illinois

Library of Congress Cataloging-In-Publication Data

Petersen Dan.
Measurement of safety performance / Dan Petersen.
p. cm.
Includes bibliographical references and index.
ISBN 1-885581-49-1 (alk. paper)
1. Industrial safety I. American Society of Safety Engineers. II. Title.

T55.P355 2005
620-8′6--dc22

2004026964

Managing Editor: Michael Burditt, ASSE
Project Editor: Jeri Ann Stucka, ASSE
Copyediting, Text Design, and Composition: Cathy Lombardi
Cover Design: Publication Design

ISBN 1-885581-49-1

Printed in the United States of America on acid-free, recycled paper.

10 9 8 7 6 5 4 3 2 1

Table of Contents

Introduction

THE MEASUREMENT OF SAFETY PERFORMANCE IS, I believe, industry's most serious problem, and it has been a stumbling block for many years. The measures we have used traditionally are often not reliable and thus invalid. This book is about attempting to find some answers for this problem.

In the early days of safety, we did not worry much about measurement. It was not until 1926 that the National Safety Council began to keep records, and they chose to measure safety performance at all levels (from first-line supervisor to national statistics) by counting the number of accidents that occurred. They used an ANSI standard for this until OSHA arrived and instituted the OSHA incident rate.

For years we have been uncomfortable with these results measures, but we continued to use them.

This book is about alternatives to those measures. Its early chapters discuss the weaknesses of the traditional metrics and the reasons for needing some different approaches. In later chapters some criteria are established for measures. Chapter 5 looks at measuring individual managers' performance, as do Chapters 6 and 7, which define the real key to measuring at the micro level; and measuring system improvement is discussed in Chapter 8, which looks at system elements and what metrics are appropriate for them. Results measures are covered in Chapters 9 and 10, and national results in safety and the measures needed there are broached in Chapter 11. Three appendices provide different measures that can be used, and a fourth offers examples of how some organizations arrive at their measures.

There are some things this book is not intended to be: It is not a book on research design. It is not a book on sampling methods. It is not a book on statistical analysis. It is not a book on Statistical Process Control. It is not a book on measuring level of risk. It is not a book for academicians.

It is a book for safety professionals who need some ways to measure the performance of their organization. It is a book for executives who would like to know how well their organization is doing, who would like to know if things are under control.

Most managers want to know how well they are performing. Are their safety systems working? Are they in control? They are interested in measures that will tell them that. Answers to such questions are most likely to come from management concepts, Deming philosophies, and similar avenues, instead of from number crunching, so our thrust here is to concentrate on activity measures and perceptions and to do less traditional massaging of the injury data.

Our search seems to have led us to look at what management is doing every day to prevent loss rather than at how the injury figures are stacking up. More importantly, we want to determine if management is doing the right things—the things that will get results—and who is best equipped to reach those conclusions.

Dan Petersen

CHAPTER 1
The Weaknesses in Safety's Measurement System

MEASUREMENT HAS BEEN A PROBLEM in the safety field for years. Despite analyzing the standards over and over again, it has been difficult to come up with meaningful measures of safety performance. Organizational Resources Counselors (ORC) in Washington, D.C., formed a task force of top safety professionals to study the whole issue. They spent over two years in monthly meetings discussing the inherent difficulties.

Steve Newell, a consultant with Organizational Resources Counselors, acknowledges that the current system for OSHA reporting was developed in 1970 to give the U.S. government a snapshot of occupational health and injuries in the workplace. Newell, who wrote the blue book used by safety and health professionals throughout the country, points out that those guidelines were never intended to give an accurate picture of safety at the plant level. "We were trying to cast a broad net to see what was out there," he says of his efforts fifteen years ago when he worked at the Bureau of Labor Statistics. "We realized afterward that people were using the numbers at the facility level." (Conley, 2000)

OSHA recordables, according to the heading of the OSHA 300 log, include every occupational death, every nonfatal occupational illness, and those nonfatal occupational injuries that involve one or more of the following: loss of consciousness, restriction of work or motion, transfer to another job, or medical treatment (other than first aid).

Complicating the measurement issue are employees who use their own discretion in reporting incidents to OSHA, explains Thomas Durbin, director of safety and workers' compensation for Pittsburgh-based PPG Industries, Inc., whose company manufactures coatings, chemicals, flat glass, and fiberglass. "If you are looking at disciplinary time off or a serious reprimand because you failed to follow a safety rule, and have an injury that can be taken care of by your personal physician, you may not choose to report it," he suggests. Also, if co-workers will lose out on a safety-related bonus because one employee gets injured, that employee will feel tremendous pressure not to report the incident.

In addition, Dee Woodhull, also a consultant for ORC, relates how OSHA's own criteria often require reporting incidents that have nothing to do with a safety program. "If you get a chicken bone stuck in your throat while you're eating lunch in the cafeteria, that's recordable," she says. Many companies have come to believe the numbers "are not representative of whether people are actually being hurt." (Conley, 2000)

CREATING FORWARD-LOOKING MEASUREMENTS

Safety metrics are important for several reasons. As Thomas Durbin says, "You need some measure of your performance. You need to be able to track your progress either up or down. It also is important to have some measures that will tell you what you need to change in order to impact performance." (Conley, 2000)

Even if they are accurate, OSHA recordables might not be the best safety metrics. "If you drive your safety program just based on OSHA recordables, you won't get anywhere because you are measuring the after-the-fact result," says Joe Holtshouser, Global Team Leader of Health, Safety, and Regulatory Affairs for Goodyear Tire and Rubber in Akron, Ohio. You are not going to gain anything by just measuring your failures. While such measures help point out larger flaws in a safety program, he adds that they are not sensitive enough to highlight specific changes that might lead to improvements. "You have to get ahead of the process and look at leading indicators," says Holtshouser. *A leading indicator is predictive.* It measures what people are doing today that might prevent illness or injury tomorrow, in contrast to *a trailing indicator which measures the results of lapses in the safety program.*

In addition to OSHA recordables, trailing indicators include the costs associated with:

- disability
- workers' compensation
- liability and litigation
- regulatory citations or penalties.

Meanwhile, leading indicators often focus on actions taken, such as:

- a process hazard review
- an incident investigation
- behavioral observation
- safety audits
- employee attitude surveys
- training records
- measures of the potential for incidents.

Newell jokes that, perhaps as penance for writing the blue book, he has been leading an effort since August 1998 to create a forward-looking system for measuring safety.

Dee Woodhull explains that an integral part of that effort is to "understand the failures in the safety management system that allow incidents to happen, so that action can be taken to prevent them before people are injured." Woodhull believes that safety programs can be more effective if you can understand the processes that lead to the failures. "These processes involve the entire organization, not just the safety program, so it is important to drill down into the root causes of incidents by looking at management systems. Metrics must be designed to link to the cause and effect relationships within the management system in order to drive performance improvement." (Conley, 2000)

Perhaps our inability to create these needed measures is one reason for our lack of excellent safety performance. Having perceived the importance of measurement in obtaining safety performance, we then hit our biggest snag: *What should we measure?*

Measuring Failures

Should our failures be measured by accidents that have occurred in the past? If this is in fact a good measure, as historically we have believed (for that is what we usually measure), then what *level* of failure should be measured? One level of failure to measure is *fatalities*. These are used to rate national highway traffic safety endeavors. Is the measure of fatalities, then, a "good" measure? Obviously, we cannot answer this question until we examine the size of the unit being measured. Fatalities could be a good measure for assessing the national traffic safety picture, but would be ridiculous to use in rating a supervisor of ten factory workers. Such a supervisor could do absolutely nothing to promote safety and still never experience a fatality in his or her department. Obviously, measuring fatalities would make little sense in this case.

Unfortunately, the traditional *frequency rate* is not much better when it is used to assess supervisory performance in safety. It measures a level of failure somewhat less than a fatality (an injury serious enough to result in a specified amount of time lost from work), but the fact remains that a supervisor of ten workers can do absolutely nothing for a year and attain a zero frequency rate with only a small bit of luck. Rewarding such a supervisor actually reinforces nonperformance in safety. He or she learns that you can do nothing and still get a reward. While this may sound a little ridiculous, it does describe what is going on in many safety programs today.

Tailoring Performance Measures

If fatalities or frequency rates are poor measures of supervisory performance, what is a good measure? Or, more importantly, what is wrong with using fatalities or frequency rates as a gauge? Perhaps measuring our failures is not the best approach to judging safety performance. After all, this is not the way people are judged in other aspects of their jobs. We do not, for instance, measure line managers by the number of parts employees in their departments failed to make yesterday. And we do not measure the worth of salespeople by the number of sales they did not make. Rather, in cases like these we decide what performances we want, and then we measure to see whether we are getting them.

So what would be a good measure of supervisory safety performance? Perhaps the better question is: What set of criteria can we develop for measuring supervisory safety performance? Or the safety performance of the corporation? Or our national traffic safety performance? Or anything else related to safety?

Even a brief look at the problem of measurement shows us that we need different measures for different levels within an organization, for different functions being performed, and perhaps even for different managers. What might be a good measure for one supervisor of ten people may not be a good measure for another supervisor of ten people, much less for a plant superintendent or the general manager of seven plants and 10,000 people. What might be a good measure for the supervisor of a foundry cleaning room may be inappropriate to use in judging how effectively a supervisor follows OSHA rules.

TRADITIONAL MEASURES

For a number of reasons, safety has used one traditional measure for performance at the national, state, organization, location, department, and even supervisor or team level—the *OSHA incident rate*. This rate measures the number of incidents (of a defined degree of resultant severity) per 200,000 man-hours worked in a facility, unit, company, state, or country. The rate was an offshoot of a pre–OSHA American National Standards Institute (ANSI) guideline (Z16.1) that attempted to measure the same thing with different definitions of an incident on a one million man-hour base.

There are serious flaws in using these "results" indicators:

1. They have little statistical validity in smaller units, measuring mostly luck, not performance.

2. They don't really tell most companies if they are improving—whether their systems are better.

3. The measures are not diagnostic. They do not suggest why an organization is performing better or worse.

4. The measures do not tell an organization what it needs to do to fix what's wrong to make the organization more effective.

Newell, who wrote the OSHA rulebook, today points out the following problems with the current measures:

- Relying on any single metric is problematic.
- OSHA rates do not drive superior safety and health performance because they are overly inclusive and not very accurate; in fact, they are less and less accurate the more pressure you put on them.
- The safety and health (S & H) measurement mindset is one of tracking failure or showing loss avoidance, not one of positive contribution to the business.
- S & H metrics undermine management credibility.

Newell suggests a "modified use of OSHA data": Continue to record and report according to the regulatory requirements, but verify data and identify a subset of cases for measurement and accountability that are reasonably serious and connected to the workplace. (Newell, 2000)

There is clearly a strong need for metrics that provide better data to managers who are attempting to curtail injuries in their facilities.

HOW METRICS AND THE SAFETY MANAGEMENT FIELD HAVE EVOLVED

When safety efforts started there were no real measures, there were only gross estimates. National Safety Council (NSC) estimates indicated that between 1912 and 1996 unintentional work deaths per 100,000 people were reduced by 90 percent, from 21 to 2. They estimated that 18,000–21,000 workers' lives were lost in 1912 while there were only 4,800 work deaths in 1996 in a workforce more than triple the size and producing thirteen times the goods and services.

Then in the early years of the profession we began using accident frequency and severity rates governed by ANSI Standard Z16.1. The National Safety Council used that standard from 1926 until about 1972 when they adopted the OSHA incidence rate. The following OSHA measures are currently in use (for 100 full-time employees per year, or 200,000 man-hours worked):

- total cases
- nonfatal cases without lost workdays
- total lost-workday cases
- cases with days away from work
- a measure of fatalities.

Beginning with the 1992 data-year, the NSC adopted the Bureau of Labor Statistics' Census of Fatal Occupational Injuries (CFOI) figure as the authoritative count for all work-related deaths.

NONACCIDENT MEASURES

The other measure under general industrial use is the *audit*: a predetermination of what should be done in a safety system, followed by periodic checking to determine whether it is in fact being done.

Over the past forty years, safety professionals have created many types of audits: checklists, yes-no audits, detailed audits, quantified audits, and audits that end up with scores or points awarded. Firms have built these tools internally or purchased them externally, then sent out individuals or teams to perform the audit. Some have questioned the validity of accepting audits as a measure of excellence, unless these audits have passed some rigorous tests.

Initially, little effort was made to correlate audit results with the firm's accident record. When such studies were performed, results were surprising, often showing a zero correlation. This was true in the 1990s at ESSO Resources (Imperial Oil) in Calgary, Canada; Venture Stores, headquartered in St. Louis, Missouri; and at Frito-Lay Corporation of Plano, Texas.

ARE WE GETTING AN EFFECTIVE MEASURE?

Whether an audit is effective may depend on whether results are correlated with accident records in numbers large enough to show validity. This can be computed easily using statistical formulas, or from scatter diagrams derived via the statistical process-control technique. (Audits are discussed in some detail in later chapters and in the appendices.)

In recent years, organizations have also been using perception surveys (also discussed with more depth later in this book) to gain insight into what is promoting or hindering good safety performance.

The evolution of the safety metric has been simplistic, concentrating almost entirely on results measures, downstream measures, and often on measures with little statistical validity.

Results measures leave us with an uncertainty about what they mean, or even which ones to look at, and different measures show different trends. For instance, according to the Bureau of Labor Statistics, in a recent 25-year period, the United States showed a reduction of:

- 0.17 percent per year in total cases
- only 0.02 percent per year in cases with days away from work.

(This measure is probably closest to the pre-OSHA measure, ANSI Z16.1.)

What this all means is that even when we can use these incident statistics (when they become somewhat valid and reliable because of the large database), perhaps they show us considerably less progress than we originally anticipated.

References

Conley, M. "How Do You Spell Effectiveness?" *Safety and Health*, June 2000.

Grimaldi, J. and R. Simonds. *Safety Management*. Homewood, IL: Richard D. Irwin, 1975.

National Safety Council. *Injury Facts*. Itasca, IL: NSC, 2001.

Newell, S. "Safety and Health Performance Metrics: Results from the ORC Alternative Metrics Task Force." Washington, D.C.: Organizational Resources Counselors, October 2000.

Petersen, D. *Techniques of Safety Management*. Des Plaines, IL: ASSE, 2003.

CHAPTER 2

Different Measures for Different Levels

YOU MAY NOTICE THAT THIS BOOK uses certain terms and titles to designate management levels; "supervisor" or "first-line supervisor" designates the level of management closest to the actual work group. Some companies today do not use these terms, preferring "team leader" or "team coordinator." We use the term "middle manager" to refer to the person to whom the supervisor or team leader reports. This person could also be referred to as a "manager," a "superintendent," or a "plant manager."

While there are a number of management levels in any organization, there are two general levels on which to measure safety performance: micro and macro. Micro measures are used at the lower, smaller units of an organization. Macro measures are used for larger units, or an entire organization. For instance, a macro measure could be the incident rate in an organization, while a micro measure could be the number of inspections, contacts, or observations made by a supervisor.

MACRO MEASURES

For many years accident measures like the number of accidents, frequency rates, severity rates, and dollar costs were used to measure the progress of the organizational unit because practitioners felt comfortable using them. These *results measures* did not reveal whether the overall safety system was effective, diagnose what was or was not working, or indicate whether the system was in or out of control.

Although it long ago became clear that these measures offer little helpful data, they continue to be used today, perhaps for the following reasons:

- OSHA requires firms to implement these measures.
- At times, compliance directions are dictated by these measures.
- The National Safety Council publishes these measures regularly, as does the government.
- Some industry groups use them to compare member companies.
- Most writers quote them.
- Most companies use them internally to judge safety-system effectiveness.

Often, however, they promote some questionable activities, such as:

- setting a goal to reduce injury rates from 3.0 to 2.0 (or even from zero to zero)
- replacing a manager who does not reach this goal
- deciding who is "good" and "bad" in order to determine who should receive an inspection or audit
- determining which company is "best" within an industry, or which location is "best" within a company.

Are these accident data useful for anything? It would save much time, effort, and money simply to answer "no" and focus on meaningful measures. However, it's not that easy. Why? Because many safety professionals do not agree that they are useless, and most executives object to losing these statistics. Despite the objections, it would be wise for practitioners to wean their companies away from dependency on such figures.

If it is true that these results measures are ingrained in most safety programs, that most executives believe they mean something, and that OSHA requires them, then why should we consider other measures?

1. Because "results" often measure luck rather than the steps taken to reduce injuries. One supervisor of ten people can do nothing and still have a zero injury record while another concerned

supervisor may have injured employees regardless of what he or she has done; this is the "luck" factor. The lower an organization's results measures are, the more these become an inadequate measure of actual performance of the safety system.

2. Because these measures do not really discriminate between poor and good performers.
3. Because results measures do not diagnose problems.
4. Because they are grossly unfair if used to judge individual managerial or supervisory performance.

There is no disputing that measurements are needed on the macro level for the following reasons:

1. To determine the effectiveness of our safety and health efforts:
   a. Is the system better today than yesterday?
   b. Which elements of the system are working and which are not?
   c. Which units are getting results and why?
   d. Where should we place our efforts next year?
2. To demonstrate the value of our safety and health efforts:
   a. Which components are paying off?
   b. Which components are of no value?
3. To provide cost-benefit analyses of the safety programs to top management.
4. To sell management on a new project for the safety program.
5. To discover why safety programs should be maintained in the future or why they should be eliminated.

What Results Tell Upper Management

Results measures reflect failures based either on incident counts or nonincident outcomes, but the validity and reliability of measuring failures depend greatly on the size of the unit in question. Before-the-fact results measures (used to judge a middle manager's performance) can include safety sampling, inspections conducted by upper management, and rating subordinate performance. Since failure measures can be used a bit more at the top management level, traditional safety indicators are a little more useful.

FAILURE MEASURES

These measures are generated from the injury record-keeping system. To measure upper-management safety performance, injury records should:

- be broken down by unit
- provide insight into the nature and causes of accidents, particularly system-level failures
- examine the causes of human error, particularly at the system level.

Beyond these broad guidelines, each firm should devise a macro-level measurement method to meet its specific needs. An important criterion that might be included in such a method is one established by Edwin Zebroski to determine the likelihood of catastrophic events.

MICRO MEASURES

Measurements at the micro level are used to drive management or team performance: "What gets measured and rewarded gets done."

It is important for these measures to be valid; that is, they must indicate whether activities have actually been performed, and these measures must flow into the daily numbers game, as well as the overall performance appraisal system.

Activity Measures: Supervisor or Workgroup

This examination of supervisory measurement looks at what the supervisor *should do* to obtain results and determine whether the supervisor *actually takes these actions.* These activity measures have certain distinct advantages:

- These measures are flexible and allow for individual supervisory styles. The same measure need not be used for all supervisors.
- They are excellent for use in objective-setting approaches.
- Feedback is swift since most of these techniques require supervisors to report their level of performance to the boss. (They are self-monitoring.)
- They measure the presence rather than the absence of safety.
- They are usually simple and therefore administratively feasible.

Activity measures are the most valid of all measures in determining supervisor performance.

BEFORE-THE-FACT MEASURES

These measures assess supervisor or workgroup action taken *before* accidents might occur. For example, a supervisor's work area is periodically inspected to measure how well he or she is maintaining physical conditions. This determines whether there are any dangerous conditions and, if so, how many things are wrong.

To measure how well a supervisor communicates with his or her employees, their work behavior can be measured (safety sampling). Safety sampling measures the effectiveness of safety efforts in curtailing accidents. It involves taking periodic readings of how safely employees are working. Like all good accountability systems or measurement tools, safety sampling is motivational.

Activity Measures: Middle Management

At the management level we often find both results measures and activity measures. Although results measures often receive greater emphasis at the middle-management level, some activity measures should be retained. At this level, activity measures are simple gauges of how middle managers perform necessary tasks.

Middle-management activities, however, should be quite different from supervisory activities, and this requires making some changes in the measures used. Normally, middle managers strive to motivate their subordinates (supervisors) to take action regarding safety. Thus, middle managers can be measured based on whether they meet with their supervisors and monitor the quality of their work. They may also be judged by whether their subordinate managers are in fact doing the work, which can be measured by a roll-up of subordinates' performance.

DEFINING THE SHAPE OF A SAFETY SYSTEM

What criteria should a safety system meet? The following are examples of six criteria used by a number of companies:

1. A system must be in place that ensures daily proactive supervisory or team activities.
2. The system must actively ensure that middle-management tasks and activities are carried out in three areas:
   a. ensuring regular subordinate performance (supervisory or team)
   b. ensuring the quality of that performance
   c. engaging in certain well-defined activities to show that safety is so important that even upper managers are doing something about it.
3. Top management must be visibly demonstrating that safety has a high value in the organization.
4. Any worker who chooses should be actively engaged in meaningful activities that are safety-related.
5. The safety system must be flexible, allowing choices of activities at all levels to instill ownership.
6. The safety effort must be seen as positive by the workforce.

Measurement is an integral and most important first step to take in judging the effectiveness of a company's safety system. Measures indicate how well the organization is performing today. The measures can take many forms, including incident statistics, behavior sampling (used as a metric), in-depth worker interviews, and perception surveys.

Perhaps the best reason for measuring results in safety is to establish a starting point for effecting change and improvement in an organization. That process of change is threefold:

1. Defining reality: Where are we today?
2. Defining the vision: Where do we want to be?
3. Defining the path: How do we get there?

References

Zebroski, E. L. "Lessons Learned from Man-Made Catastrophes," in *Risk Management: Expanding Horizons in Nuclear Power and Other Industries* (Ronald A. Knief, ed.). New York, NY: Hemisphere Publishing Corp., 1991.

CHAPTER 3
What Should We Measure?

IN CHAPTER 2 WE DISCUSSED macro and micro measures, results and activity measures, and performance measures for supervisors and mid-level managers. We also introduced the terms "failure measures" and "before-the-fact measures."

These are terms often used in the current safety environment. We also hear about "upstream" and "downstream" as well as "leading" and "trailing" indicators. While confusing, all of these terms, garnered from different industries, are simply different ways of trying to describe how we measure what we are doing to prevent or control unsafe acts or conditions; they are the framework from which we judge whether or not we are succeeding.

For the remainder of this chapter, we'll concentrate on two terms:

- Activity measures: What steps do we take to ensure safety?
- Results measures: Have we accomplished those safety objectives?

BUILDING THE MEASUREMENT FRAMEWORK

It is important at the outset to define the purpose for taking measurements. Perhaps the first question to ask is: What elements should we measure to determine our safety record? In attempting to answer this simple question, we must examine a number of things:

- We must determine which level of the organization we are measuring. At the supervisor level, we measure to motivate him or her to carry out the activities defined for or by that person. At the unit level (location, plant, corporate level), we measure to determine the effectiveness of our efforts and systems. This means that the measures will be different for each unit.

- We must focus on the current measurement terms we are using, such as leading and trailing indicators, upstream and downstream measures, activity and results measures.
- We must ascertain which measures are available to us, and which ones will actually drive performance and results.

Two categories of available tools are (1) activity measures and (2) results measures. As noted in Chapter 2, activity measures should be used at the supervisory level; activity measures, along with some results measures, should be used at the middle- and upper-management levels; but the pure results measures should be reserved for the executive level. With rare exceptions (like safety sampling), this rule of thumb gives us statistical validity and is extremely important (see Exhibits 3.1 and 3.2).

As indicated, we can use either activity measures or results measures to determine performance, and we can use them at the supervisory, managerial, or systemwide level, provided we exercise caution in the measurement selection. At the supervisory level, activity measures are the most appropriate, whether counting the number of inspections made, the number of people trained, or the number of observations made. These activity measures are equally appropriate at the managerial level, and can even be valuable at the system level through the use of audits or questionnaires. All activity measures are extremely well-suited to management by objectives (MBO) or safety by objectives (SBO): Were the objectives reached? Using these measures in SBO assumes that the original objectives were well written (realistic, highly specific, measurable, and under the direct control of the objective-setter).

The traditional safety measures, such as frequency rate or severity rate, should not be used at the lower levels except over long periods of time and probably only as a quality check.

■ Exhibit 3.1 **Activity versus results measures**

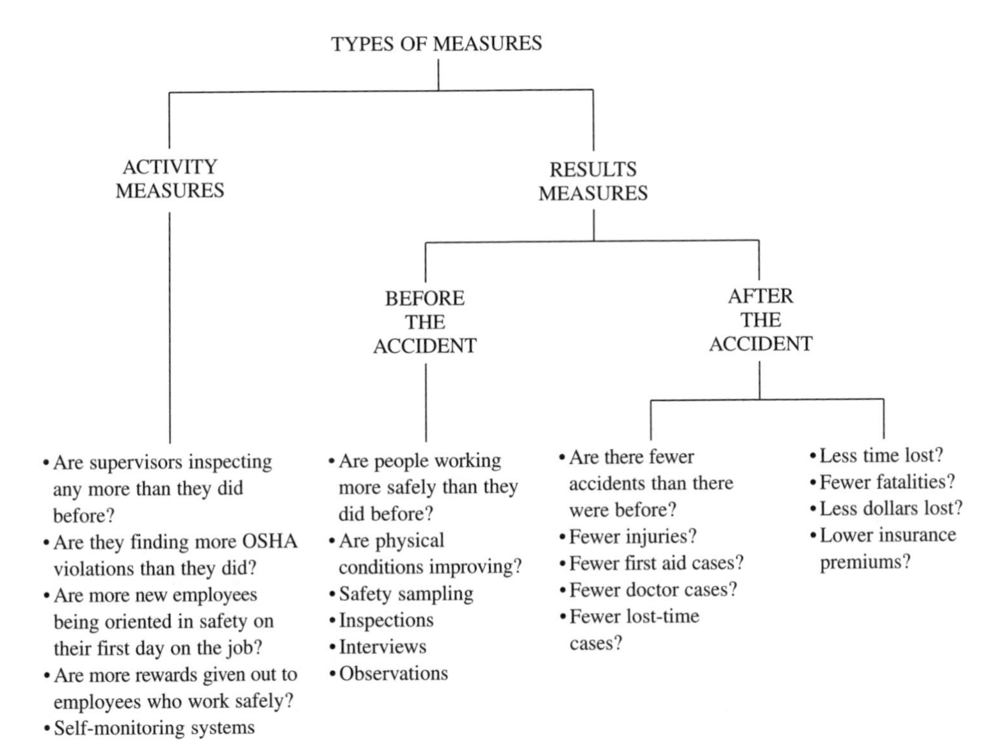

Activity Measures at the Supervisory Level

The examination of supervisory measurement begins by looking at what tasks the supervisor performs to get results and determining whether the supervisor actually is doing them.

Examples of activity measures at the supervisory level are shown in the left column of Exhibit 3.2 . They include the number of

- inspections performed
- employees trained
- hazard hunts completed
- observations made.

■ Exhibit 3.2 **Activity and results measures for supervisors, managers, and systemwide safety programs**

ACTIVITY		
SUPERVISOR	**MANAGER**	**SYSTEMWIDE**
Objectives Met	Objectives Met	Audit Vehicles
# Inspections	Use of Media	Questionnaires
# Quality Investigations	# Job Safety Analyses	Interviews
# People Trained	# Job Safety Observations	
# Hazard Hunts	# One-on-Ones	
# Observations	# Positive Reinforcements	
# Quality Circles	Group Involvement	
RESULTS		
SUPERVISOR	**MANAGER**	**SYSTEMWIDE**
Safety Sampling	Safety Sampling	Safety Sampling
Inspection Results	Inspection Results	# First Aid or Frequency
	Estimated Costs	# Near Misses or Frequency
	Control Charts	Property Damage
	Property Damage	Estimated Cost Control Charts

Activity measures weigh the results of supervisor action before an accident occurs. For example, a periodic inspection is made of a supervisor's work area to measure how well he or she is maintaining physical conditions. This is a measure of whether things are wrong and, if so, how many things are wrong.

We can also measure how well a supervisor gets through to the people in the department by measuring employees' work behavior (e.g., safety sampling). One recent and excellent measurement tool is the perception survey.

Activity Measures at the Middle-Management Level

While results measures are emphasized more at the middle-management level, some activity measures are valuable here as well. At this level, these are simply measures of the task performance attained by middle managers. Thus, everything we have said about supervisory performance measures applies also at the middle-management level.

However, supervisory responsibilities and middle-management responsibilities entail quite different tasks, necessitating alterations in the measures in play at this level.

First of all, middle management plays a threefold role:

1. ensuring subordinate supervisory performance
2. ensuring the quality of that performance
3. engaging in activities that show all reporting personnel that safety is so important that each employee should do something personally about it.

Normally, we want middle managers to motivate their subordinates (supervisors) to accomplish some safety objectives. This allows us to measure whether they meet with their supervisors, check on these supervisors, or monitor the quality of the supervisors' work.

Results Measures at the Upper-Management Level

At the top management level, we concentrate on results measures, but we can work with both activity measures and results measures. Here our traditional indicators become more useful. The validity and reliability of our results measures, however, may depend upon the size of the unit with which we are working.

AUDITS

The audit of safety performance provides a measurement of activities by upper- and middle-management levels (management of a location, for instance). Such an audit may be problematic if an organization thinks that its results are predictive of the incident record.

MEASUREMENTS TO USE

The literature on measurement, in general, and on measuring safety, in particular, suggests the following:

1. Quit looking at accident-based measurements to assess system effectiveness.

2. Use audit results only when certain a correlation exists between the audit and safety results over time and across large numbers.

3. Use properly constructed perception surveys as primary measurement and diagnostic tools (downstream or trailing indicators).

4. Use behavior sampling and activity goals that have been met on an ongoing basis by small units as primary motivational measurement tools.

Andrew Hopkins (ORC) suggests that we use process (activity) indicators rather than outcome indicators:

> ... indicators which measure safety-relevant processes rather than outcomes such as injury or fatality rates. Process indicators will differ from workplace to workplace depending on just what processes are relevant. Moreover, they must measure things which occur with reasonable frequency so that variations have a chance of being statistically significant and hence indicators of real change in safety performance. For example, if in a certain workplace hoses left unrolled result occasionally in workers tripping, and on rare occasions in a lost time injury, then counting the number of hoses left unrolled at any one time is likely to be a useful process indicator. If this count is repeated at randomly scheduled times, trend data can be rapidly built up.

It is important to distinguish between two types of process indicators: those which focus on the behavior of employees, and those which measure management activity. Consider, first, indicators of employee behavior. Examples would include: the percentage of people wearing personal protective equipment, e.g., hearing protectors, at required times; the frequency with which danger tags are being used as required; and measures of good housekeeping, such as rolling up hoses. One of the best features of such indicators is that merely publicizing the data within the workplace focuses attention on the problem and is likely to lead to safety improvements without the need for more direct or punitive management intervention. Moreover, the use of these indicators has the effect of involving people in the task of improving safety and creating a culture of safety. Such indicators have the following advantages:

1. They are sensitive indications of health and safety performance, enabling a workforce to detect whether safety is improving in a matter of weeks rather than months or years.
2. They are positive, focusing on how good rather than how poor safety is.
   (1) They are a direct measure of safety performance, focusing on how well personnel are complying with their own agreed safety responsibilities.
   (2) The results can be used as a powerful performance feedback.
   (3) They involve all workers and achieve "ownership."

There is however a significant drawback to such indicators. They are focused on and aimed at changing the behavior of employees, not managers. Yet it is managers who are ultimately responsible for health and safety, and who are in the best position to take action on such matters. Hence the importance of indicators which measure the safety-related activity of management. Examples here might include, depending on circumstances: percentage of workforce which has received safety training, or percentage of safety audits which have been completed on schedule. The general principle should be to have management specify its safety management plans and procedures and then to construct measures which assess how well these are being carried through in practice.

References

Hopkins, A. "The Limits of Lost Time Injury Frequency Rates," in *Positive Performance Indicators Beyond Lost Time Injuries*. Australia: Worksafe, 1999.

Organizational Resources Counselors. *Occupational Safety & Health Metrics Guide*. Washington, D. C.: ORC, 2003.

CHAPTER 4

Choosing Criteria for Safety Measurement

THE DEVELOPMENT OF CRITERIA for good safety measures is certainly not an easy task. We have been grappling with it for years, and the most noted theorists and scholars in safety have been writing on the problem since the late 1950s.

Roy Green suggests:

> Ideally, to be effective, the approach chosen to measure the performance of firms and organizations should fulfill [sic] three basic requirements. To begin with, it must be sufficiently accurate to capture all the elements of performance at the workplace, including safety performance. Second, it must be sufficiently simple to be understood and acted upon by managers and employees responsible for improving performance, and, finally, it must be sufficiently dynamic to encourage as well as measure the continuing process of organizational change and improvement. (Green, 1994)

We need more than a list of general criteria, however. For instance, such a list usually includes an item called *statistical reliability*, which has to do with whether a measure tends to fluctuate wildly when there has not been much change in the system being measured. The criterion of statistical reliability would obviously be useless for measuring a first-line supervisor's safety performance. A supervisor who shows no safety improvement (zero performance level) could have a perfect record or a miserable record. At the supervisory level statistical reliability is next to impossible to achieve with any measure; the database is simply too small.

Other examples could be given, but perhaps none are necessary. Different criteria, and therefore different measures, seem to be needed for different organizational levels and perhaps for different functions.

This chapter describes criteria for different levels and different functions.

Safety-Measurement Criteria for Employees

Few measures of safety performance are used at the lowest organizational level, that of the employee. Those that are used are essentially individual assessments of performance. Nonetheless, we can consider a few possible criteria for a measure of employee safety performance:

1. It should be constructed so that it can be used to affect an employee's rewards (appraisals, promotions, bonuses, etc.).
2. It should be constructed in such a way that it recognizes or can be used to recognize safe performance (rather than unsafe performance).
3. If possible, it should be self-monitoring. That is, the employee can record achievements himself, thus reinforcing his behavior and his success toward achieving a goal.
4. It should be motivating to the employee. Goal achievement is one of the best motivators.

Typically, the measures used at this level, regardless of whether the employee has had any injuries, might meet criteria 1 and 2, perhaps 3, but probably not 4 (unless there is negative reinforcement).

Safety-Measurement Criteria for First-Line Supervisors

Measurement is more crucial at the supervisory level, and the measure (which is the motivator here) must accomplish more than at the employee level. In addition to the criteria for measurement at the employee level (specifically, the self-monitoring and reward-recognition aspects), the following criteria should be used:

1. It should be flexible enough to encompass individual managerial styles and different strategies supervisors use to get things done.
2. It should give swift and constant feedback.
3. It should provide the ability to judge promotability.
4. It should get the supervisor's attention.
5. It should measure the presence of safety activity, not just its absence (as indicated by accidents).
6. It should be sensitive enough to indicate when effort has slowed.
7. It should provide an alert, showing that something is wrong.
8. It should be understandable to those at all management levels.
9. It should allow for creativity.
10. It should be valid; that is, it should measure what it was intended to measure. If you want supervisors to perform accident investigations, inspections, and training, the measure should show whether you are getting these performances. It should not measure only failures (accidents) as an indication of whether you are getting the desired performance.
11. It should be mainly activity-oriented.
12. It should be meaningful.

By definition, all of the safety-measurement criteria listed above are positive, none is negative. Obviously, no one measure will meet every one of them. Thus, there will have to be some trade-offs, and it will be necessary to devise and use measures that meet as many of the criteria as possible. Often, at this level, measures are the OSHA incident rates, which meet almost none of our listed guidelines. How many are viable standards will depend on the organization and its systems. Each organization can assess the pros and cons of each criterion and, later, of each metric under consideration.

Safety-Measurement Criteria for Middle Management

The same criteria apply at the upper levels of management as at the lower levels, although some new ones become important. As indicators of performance at the middle-management level, the measurement criteria should be sensitive to change (able to alert management to new situations and problems). As opposed to the supervisory criteria, the measurements for middle management should be both activity- and results-oriented.

Typically, the OSHA incident rates, which meet almost none of these criteria, are the institutionalized measures.

Safety-Measurement Criteria for Upper Management

The measure used at this level indicates how well the company or unit is doing. It is used exclusively inside the organization. It can indicate the progress of the entire organization and could be construed as constituting a judgment of the president and controlling officers. Such a measure is not intended for use in comparing the performance of one corporation with that of another; different criteria will be suggested for this type of measure. All of the criteria that apply to middle management also apply to upper management. Additional criteria for upper management are:

1. It should be statistically reliable; it should not fluctuate without reason.
2. It should be objective.
3. It should be quantifiable.
4. It should be stable.
5. It should ensure input integrity.
6. It should be primarily results-oriented.
7. It should be computerized easily.
8. It should point out weaknesses in the system and thus make it possible to take preventive action.

Note that as the management level increases, the measurement orientation changes: at the supervisory level, it is mainly activity-oriented; for middle management, it is a combination of activities and results; and now at the upper-management level, it is primarily results-oriented.

Since upper management considers larger numbers of people, we can demand certain things of our measures that we could not expect at other levels. These are ideal criteria to meet, and this is the appropriate level of the organization to provide such measurements. Again, the standard here is usually the OSHA incident rate, which might meet only the quantification and computerization criteria.

Safety-Measurement Criteria at the National Level

Measures used to assess corporate safety performance can be used to compare one company's progress with that of another, as well as with the national average. Any measure used to judge our collective progress at the national level should be totally results-oriented and understandable to any layperson. In this case, the OSHA incident rate, which we have noted was mainly designed for quantification and computerization, does not meet those standards.

Additional Criteria

The following safety-measurement criteria should be considered for every organizational group, from employees through all the levels of management. They can even be used as rubrics at the national level.

1. It should provide a good cost-benefit ratio.
2. It should be administratively feasible.
3. It should be practical.

The criteria listed and discussed in this chapter are by no means universally accepted. However, they offer some good starting points. Unless we define our criteria, we cannot tell whether the measures we devise are worthwhile.

References

Green, R., "A Positive Role for OHS in Performance Measurement" in *Positive Performance Indicators for OHS*. Report for Worksafe Australia. Canberra, Australia: Australian Government Printing Service, 1994.

CHAPTER 5

Measuring Individual Manager Performance

THE ATTITUDE OF THE MAJORITY of managers and supervisors today lies somewhere between total acceptance and flat rejection of their role in a comprehensive accident prevention program. Most typical is the organization in which line managers do not shirk this responsibility, but do not fully accept it either, or treat it as they would any of their defined production responsibilities. In most cases their "safety hat" is worn far less often than their "production hat," "quality hat," "cost-control hat," or "methods-improvement hat." In most organizations, safety is not considered as important to the line manager as many, if not most, of the other duties performed.

On what does a manager's attitude toward safety depend? The book *Managerial Attitudes and Performance* by Lyman Porter and Edward Lawler does an excellent job of examining this, and most of what we discuss here is based on what they have said.

DESIGNING A MODEL FOR SUPERVISORY BEHAVIOR

Porter and Lawler build on motivation theory to construct a model of supervisory behavior. We will use their basic model to describe supervisory safety behavior. Motivation theory, in general, attempts to explain how behavior gets started; how it is energized, sustained, directed, and stopped. While a great number of motivation theories have been proposed, there are only two theories that have been developed to a comprehensive state. The first of these is the *drive X habit theory*, and the other is *expectancy X value theory*. The model constructed by Porter and Lawler that we are using is based on the expectancy X value theory of motivation.

Expectancy theory states that people have expectations or anticipations about future events. These take the form of beliefs concerning the likelihood that a particular act will be followed by a particular outcome. Such beliefs or expectancies could have a value between 0 (no chance) and 1 (completely sure it will follow). Stating this basic theory in simpler terms, if line supervisors expect they will get something favorable every time they perform a specific safety function, they will be quite likely to perform it.

Porter and Lawler's examination of managerial attitudes is built around the theoretical model shown in Exhibit 5.1. The model suggests that managerial performance is dependent on three primary variables: abilities, role perception, and effort expended. All are important, and managers will not turn in the kind of performance expected unless all three are taken into account.

There are two basic factors that determine how much effort employees put into their jobs: their opinion of the value of the rewards and the connection they see between their effort and those rewards. This is true of a manager's total job, as well as of any one segment of it, such as safety.

Variables in the Model

VALUE OF REWARD

This variable refers to the attractiveness of possible outcomes to individuals. If these outcomes are positive, they can be thought of as rewards. The various rewards that a person might hope to obtain are the friendship of fellow workers, a promotion, a merit salary increase, or an intrinsic feeling of accomplishment. A given potential reward means

■ Exhibit 5.1 **The management model (Porter and Lawler, 1968).**

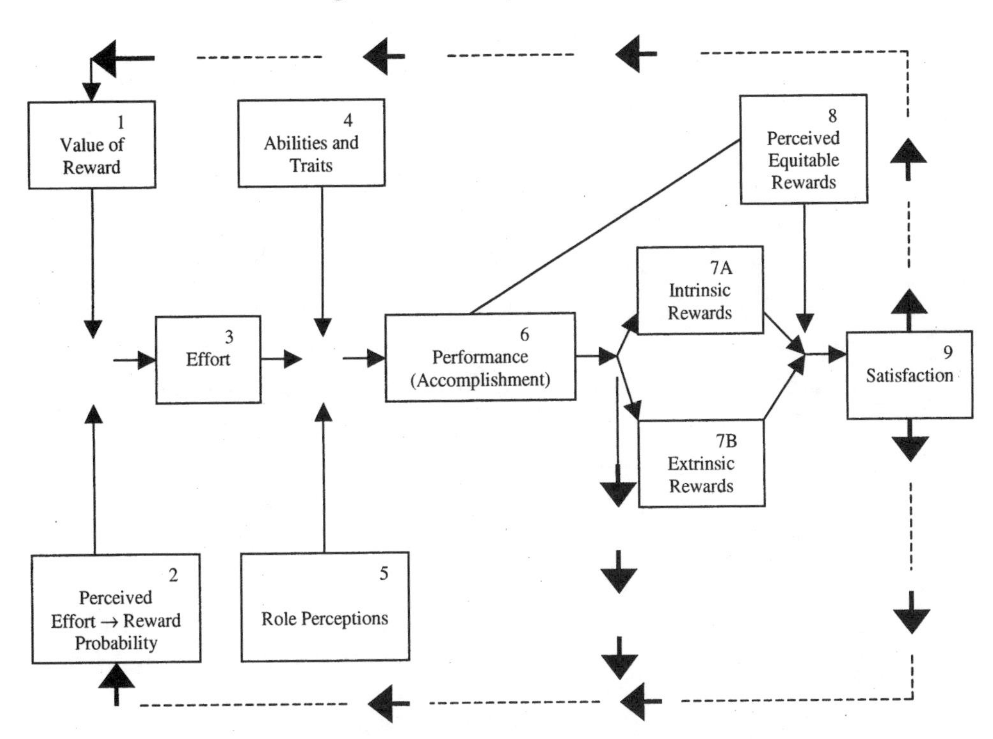

different things to different individuals. For example, the friendship of peers (i.e., workers at the same job level) might be highly desired by one worker and be unimportant to another. A promotion might have very little positive value for one person because of their lack of desire to take on increased responsibilities, but for a middle manager in a large corporation, a promotion might be a reward of extremely high value. Thus, value of reward refers to how attractive or desirable a potential outcome that results from an action is to a worker.

In safety, the manager looks at the work situation and asks, "What will be my reward if I expend effort and achieve a particular goal?" If the value of the reward management offers for achieving the goal is considered great enough, the manager will decide to expend the effort. Chances are that the manager will decide they are if the rewards are in terms of advancement and additional responsibility, rather than some of the lesser enticements that management too often selects. In safety, just as in other areas, management's chosen rewards are often too small and too unimportant to entice the line manager.

EFFORT-REWARD PROBABILITY

Effort-reward probability refers to an individual's expectations concerning the likelihood that the rewards depend on effort. Such an expectation can be divided into two parts: (1) The probability that reward depends on performance, and (2) the probability that performance depends on effort.

In safety, a manager asks the following questions when assessing how probable it is that the rewards really depend on effort:

- Will my efforts here actually obtain the results wanted or are there factors involved beyond my control? (The latter seems a distinct possibility.)
- Will I actually get that reward if I achieve the goal?
- Will management reward me better for achieving other goals?
- Will management reward (promote) another manager because of seniority, regardless of my performance?
- Is safety really that important to management or are some other areas more crucial to it right now?
- Can management effectively measure my performance in safety or can I let it slide a little without management's knowing?
- Can I show results better in safety or in some other area?

EFFORT

Effort is a key variable in this model and it should be clearly understood as being different from performance. Effort refers to whether or not the supervisor tries; performance refers to whether or not he or she succeeds.

ABILITIES AND TRAITS

Each supervisor possesses unique characteristics, personality traits, intelligence, and manual skills. Is that individual able to perform? In safety, this means ensuring, through selection and supervisory training, that the line manager has sufficient safety knowledge and abilities to control the people and the conditions under which they work. In most industries, lack of knowledge is not a problem. Usually the line manager knows far more about safety than he or she will ever need to apply. Many people agree that a manager can achieve remarkable results on accident records merely by applying management knowledge, even with little safety knowledge. If a manager does not have adequate safety knowledge, the problem is easily handled through training.

ROLE PERCEPTIONS

How a supervisor perceives his or her role is directly related to the direction of effort, including the kinds of activities the supervisor believes necessary to perform the job successfully. If the supervisor's role perceptions correspond to those of management, effort will be applied where it will count the most for successful performance, as defined by the organization. If the supervisor's perceptions are inaccurate (do not correspond to those of management), a great deal of effort may be expended without successful performance. In safety, role perception boils down to: "Does the supervisor know management's desires on accident control?" and "Does the supervisor know what the duties are?"

Analyzing role perception allows us to search for answers to some questions about our organization and about each line manager in our organization. These questions should concern, among other things, the content and effectiveness of management's policy on safety, the adequacy of supervisory training, the company safety procedures, and the systems used to fix accountability.

REWARDS

Desirable outcomes, either monetary or psychological, are provided by the supervisor (intrinsic) or by others (extrinsic). In safety, there is always an intrinsic reward, but perhaps not as often an extrinsic reward. Rewards are desirable states that a person acquires either from personal thinking or the action of others (management). For predicting future performance, the most important things to know about rewards are their perceived size and their perceived degree of connection to past performance.

PERCEIVED EQUITABLE REWARDS

A supervisor will expect a certain level or amount of reward as the result of a given level of performance. In safety, this often boils down to a simple assessment of the amount and kind of reward the supervisor thinks would be correct for safety effort. In many cases, the supervisor does not expect to receive much of a reward for promoting safety, and therefore does not give it much effort. This, however, can present a real opportunity for management. If the reward structure for safety performance can be improved to the point where the supervisor will be pleasantly surprised rather than disappointed, it will reinforce continued safety effort.

SATISFACTION

The comparison a supervisor makes between what is perceived as an equitable reward for safety performance and what is actually received (both extrinsic and intrinsic) for safety performance determines satisfaction, dissatisfaction, or pleasure with the results of their efforts. This level of satisfaction then serves to influence effort in the future.

Exhibit 5.1 shows three different feedback loops. The first of these goes from performance (accomplishment) back to the perceived effort-reward probability. What this loop says is that achieving what is set out to be achieved will help to convince the supervisor that it can be achieved. Successful accomplishment will tend to increase perception of the probability that an attempt to do it will bring success.

The second feedback loop goes from satisfaction back to perceived effort-reward probability. This loop says that if the supervisor is satisfied after comparing the reward received with the expected reward, there is a higher probability of perceived equitable rewards next time.

The third feedback loop goes from satisfaction back to the value of the reward. This says that satisfaction from the last experience reinforces the belief that the value of the reward is there. The supervisor will be "more certain" of the reward than before.

This process exemplifies the model. While it may look complex, it really is not. The model states that the determinants of performance are abilities and traits, role perception, and effort. It further states that effort is dependent on the reward and the probability that the reward is connected to effort. Upon accomplishment, the supervisor receives intrinsic and extrinsic rewards. These rewards are compared with the perception of what is equitable and determines satisfaction or dissatisfaction. The level of satisfaction influences the perception of rewards and the effort-reward probability for the future.

The model is very applicable to safety. Asking the following kinds of questions about each box in the model may best summarize this applicability:

- What are the rewards for safety performance? Compare them to the rewards offered in production, quality control, and other areas.
- Do supervisors believe this probability is high? Will attempts bring success? Will success bring rewards?
- The reward and the effort-reward probability determine the amount of effort. How much effort will your supervisors give to safety concerns?
- Abilities and traits are a function of your selection and training. What are the levels of safety performance in your company?
- Role perception is a function of safety policy and supervisors' observations. Do they perceive safety as their function?
- Performance is a function of ability, role perception, and effort. What kind of performance are you getting?

- The reward received is both internal and external. Is the external reward there? In what forms?
- Supervisors expect a certain reward for accomplishment. What is that expectation?
- Supervisors compare the reward to their expectations. Satisfaction (or lack of it) influences their efforts from this point on. What comparisons have your supervisors made?

One of the determinants of supervisory safety performance is role perception. Role perception depends first on knowledge of management's stated (and reinforced) desire that safety be a part of the supervisor's job. Management influences the supervisor's perception of the role through training, systems of fixing accountability, and policy. The process starts with policy, how it is explained in training, and how it reinforced through accountability.

Role perception then is nothing more than whether or not each supervisor perceives, or rather believes, that doing something in safety is really a part of the job. Role perception includes the interpretation of management's priorities. For example, is there belief, from management's past actions, that safety is as important as, or not as important as, production?

Accountability

Often a company, even when it writes a safety policy, does not define any procedures to fix accountability; that is, to measure the safety effectiveness in the performance of line managers. Until such procedures are designed, safety policy does not mean much. Instead of simply preaching for fifty years that the line has responsibility, we should have been devising procedures to fix such accountability. When someone is held accountable (is measured) by the boss for doing something, responsibility will be accepted. Without accountability, there is no accepted responsibility. Effort will be steered in the direction the boss is measuring because this is the role perception that is engendered by the boss's own measurement system.

The primary way that management influences the supervisor's effort is by determining the value of reward for safety performance and effort-reward probability. This is done through policy that is consistently and constantly followed up by measurements (systems to fix accountability).

Rewards for safety performance are no different from rewards for accomplishment in any other area where management asks for performance from the supervisor. While our model focuses primarily on positive rewards (i.e., peer acceptance, subordinate approval, enhancing the likelihood of promotion, merit salary increases, higher bonuses, intrinsic feelings of accomplishment, compliments from the boss), it could also include *negative rewards* (i.e., negative performance appraisals, lower raises or no salary increases, smaller bonuses, reprimands).

The main difficulty in developing a reward system is not in determining *what* the rewards will be for performance, but rather in determining *when* the reward should be given; that is, when the supervisor has *earned* it. Unfortunately, the safety field has precious few good measurement tools that can tell us when a supervisor is performing.

MOTIVATION AND MEASUREMENT

Safety work naturally suggests three ways of assessing supervisory performance: measure the activities of the line, measure the results of those activities, or use a combination of both these measures. The most used measurements seem to be based on results, which usually provide the worst data for judging performance.

First-Line Supervisors

At different levels of the organization there are major differences in what seems to motivate people. First-line supervisors have moved up a step in the organization and are in a totally different position psychologically. They used to be motivated primarily by employee peer pressure, but this is now much less important. Members of the old peer group are supervisors themselves. The group is less cohesive, and its pressure is less important. Now there is more interest in the manager and in his or her desires, because pleasing the boss means retaining the new position of power over others and perhaps being promoted again to a position of more power and greater financial gain. Thus, the first-line supervisor's primary motivational force is the boss: how he or she is measured by the boss, how to demonstrate good job performance to the boss, and how to find out what the boss really wants. However, the measurements the boss uses are not always a good indicator of what the boss really wants since many of them have been chosen by upper management. Also, priorities shift and change without any concurrent change in the measures the boss uses. Despite this, first-line supervisors strive to find out what is important to the boss, and they react to those items first.

We use the term "boss" here because it is still a meaningful term to most people in industry. We understand it is a politically incorrect term in today's world where we use "associates" for workers, "team leaders" for supervisors, and "coordinators" instead of supervisors. Regardless of today's politically correct environment, the reality is that there are still "bosses" in the world.

It is true that first-line supervisors also react to pressures from other directions. Since they must live with their subordinates, they react to them and to the pressure they exert. There is still a small amount of peer pressure, but this usually is not a strong reason for the first-line supervisor's behavior. He or she is motivated first by what the boss wants (as indicated by the measure used and by the boss's expressed wishes) and second by the members of the group. The first-line supervisor must know all these people well enough to recognize what they want and need; success on the job depends on this.

These motivational pulls are described well by a number of behavioral scientists and management theorists. The model of supervisory performance devised and tested in full by Edward Lawler and Lyman Porter is one of the most descriptive (see our discussion of this model at the beginning of the chapter).

Middle Management

What motivates middle managers is very similar to what motivates first-line supervisors, with one major exception. While they are very much interested in the performance measures their boss uses, and in the desires and wishes of the boss, middle managers are especially responsive to measures relating to the dollar—to profitability—which is so important to managerial thinking. The Lawler and Porter model is also relevant at this level, the only difference being that results measures are probably better used as indicators of performance and effort at this level than at the lower level.

John M. Cohen describes the system his organization (Winter Construction) decided on:

> Winter decided its safety performance program would measure two basic elements: *safety activities* and *safety results*. Each element is weighted to provide 50 percent of a participant's total score.
>
> The safety activities selected to be measured are those already incorporated in the company's safety and accident prevention program, including:
>
> - Regular and frequent safety inspections by the safety department,
> - Weekly site safety inspections by the superintendent,
> - Weekly safety training for hourly workers, and
> - Injury reporting and accident investigation, including separate scoring for promptness of notification and for the quality of written reports.
>
> A standard was developed for each of these activities along with corresponding point scores having a possible total of 100. The activity score is weighted at 50 percent of the total score.
>
> Winter decided that the measurement for safety results would be the incidence rate (IR) achieved by a superintendent for a calendar year measured against a target IR. The safety department, however, believed that incidence rates alone would not necessarily reflect the true caliber of a superintendent's safety performance. Many safety professionals look at incidence rates as measures of luck.
>
> A superintendent's IR is derived using the totals of work hours and work-related injuries of Winter's employees, plus those of its

subcontractors, on a superintendent's project(s). As a general contractor, Winter self-performs approximately 20 percent of its work, meaning that its subcontractors typically represent a superintendent's most significant safety management challenge. To measure only 20 percent of safety on a project would ignore a superintendent's duty to provide a safe work site for *all* workers. This method of developing an IR also serves to underscore Winter's safety responsibilities as a *controlling employer* under OSHA's *multiemployer work site* doctrine.

References

Cohen, J. "Measuring Safety Performance in Construction," *Occupational Hazards*, June 2002.

Porter, L. W. and E. E. Lawler. *Managerial Attitudes and Performance*. Homewood, IL: Irwin-Dorsey, 1968.

CHAPTER 6
Measuring the Line Manager

THE PRINCIPLE OF ACCOUNTABILITY cannot be separated from the techniques of measurement. In fact, measurements are made for the purpose of fixing accountability—without them, accountability is meaningless.

In this chapter we look at measurement techniques that fix accountability to line management. Several of the more common tools of safety measurement are discussed: inspection, accident investigation, record keeping, and statistical control techniques. These tools are examined as they relate to the principle of accountability—not just as tools in themselves. For example, investigations can be used to unearth symptoms rather than causes. Records can be used to tabulate accident types, accident agencies, injury types—not just accident causes. In the past, safety professionals tended to use inspections for the purpose of determining hazards; they have used accident investigations for the purpose of identifying an unsafe act or an unsafe condition; and they have used record keeping only to compute frequency and severity rates. This chapter discusses how we might better use these tools for what should be their primary purposes: measuring line safety performance, spotting acts as well as conditions, and unearthing causes as well as symptoms.

OVERVIEW ON LINE MEASUREMENT

Our examination of supervisory measurements begins by looking at the things a supervisor does to get results and then determining whether the supervisor actually does them. This line of thinking suggests measures like the following:

1. In the area of inspections
   a. How many have been made?
   b. How many unsafe conditions were found?
   c. How many corrections of conditions were made?
   d. How many corrections of conditions were suggested?
   e. Were behaviors observed? What types?
   f. How many poor performances were noted?
   g. How many corrections of performance were made?
   h. How many corrections of performance were suggested?
   i. Were system weaknesses found? How many?
   j. How many system changes were made?
   k. How many system changes were suggested?

2. In the area of accident investigations
   a. How many were made (in relation to the number of accidents)?
   b. Were they made on time?
   c. How many root causes were found (in the management system)?
   d. How many corrections were made?
   e. How many corrections were suggested?
   f. What was the quality of the corrections?

3. In the area of training
   a. How many new employees were trained in safety?
   b. How many old employees were trained in safety?
   c. How many five-minute safety talks were given?
   d. How many employees attended each training session?
   e. What scores were made on the tests given during training?
   f. What improvements resulted from training?

4. In the area of motivation
   a. How many employees were individually contacted?
   b. How many positive or negative reinforcements were given?
   c. How many posters were used?
   d. What other media were utilized?

5. In other areas
   a. How many job safety analyses (JSAs) were done?
   b. How many safety samples were taken?
   c. How many job safety observations (JSOs) were made?
   d. How many hazard hunts were undertaken? What did they uncover?

In short, these measures are simply counts of the supervisor's activities. They are valuable because they meet our criteria well. For instance, they are flexible and take into account individual supervisory styles, not restricting all supervisors to the same yardstick. Each supervisor can select activities and levels of performance, and then these performances and levels can be measured. Activity measures are excellent for use in the safety-by-objectives (SBO) approach because they measure the presence rather than the absence of safety. They give swift feedback, requiring supervisors to report their level of performance directly to the boss. In that sense, they are self-monitoring. Above all, they are simple, making them administratively feasible.

INSPECTION

Inspection is one of the primary tools of the safety specialist. Generally it is used to spot conditions, but seldom to spot actions. Before 1931 it was virtually the *only* safety tool, and from 1931 to 1945 it was still the tool most frequently used. Until 1960, it was the primary tool of many outside service agencies, and today it remains the primary (and sometimes the only) tool of some safety professionals.

Regardless of the organization, each person who performs an inspection should ask one key question: Why am I inspecting? The answers to that question dictate how, when, and where to inspect. For instance, if you are inspecting in order to unearth physical hazards only, you will look only at *things*. If, however, you are inspecting in order to pinpoint both physical hazards and unsafe acts, you will also look at *people*. Unfortunately, most inspections today still look only for physical hazards and not at the unsafe behavior of employees.

If the primary intent is to detect hazards that have not been seen before, the inspection will be different than if the primary interest is in checking on the inspections the department supervisor has made. If the intent is to detect hazards only, they can immediately be corrected by going directly to the maintenance department and reporting any deficiencies. If the intent is to audit the supervisor's inspection, what you find can be used to instruct and coach the supervisor so that future inspections will be improved.

Many articles have been written on safety inspections, and many of them have asked the question: *Why inspect*?

Some typical answers are:

- to check the results against the plan
- to awaken interest in safety
- to reevaluate safety standards
- to teach safety by example
- to display the supervisor's sincerity about safety
- to detect and reactivate unfinished business
- to collect data for meetings
- to note and act upon unsafe behavior trends
- to reach firsthand agreement with the responsible parties
- to improve safety standards
- to check new facilities
- to solicit the supervisor's help
- to spot unsafe conditions.

However, the most important reason for making inspections is seldom mentioned: to measure the supervisor's performance in safety.

If the line manager felt that the supervisor's performance was the primary purpose of management's inspection, perhaps he or she might do a better job of making sure that no problems could be found in the department.

If inspection is used as a measurement tool of accountability, it should prompt the line manager to inspect more often to ensure that conditions remain safe and that fewer unsafe acts are being performed.

Who Is Responsible for Inspections?

It is generally agreed that the responsibility for conditions and people resides with the line supervisor. Thus responsibility for the primary safety inspection must also be assigned to the supervisor. A *primary safety inspection* is intended to locate hazards. Therefore, any inspections performed by staff specialists should be only for the purpose of auditing the supervisor's effectiveness, and the results of those inspections become a direct measurement of his or her safety performance or effectiveness.

Historically, inspections have been important in safety. Are they still? Perception surveys conducted throughout industry pinpoint three areas that always score low; out of twenty categories rated best to worst, inspections consistently rank third from the bottom. Company employees state that only 62 percent of their companies are actually inspecting their facilities today.

Symptoms of Unsafe Conditions

An unsafe act, an unsafe condition, and an accident are all symptoms of something wrong in the management system. This should be a guiding axiom during inspections. Look behind the acts you see when inspecting, and behind the conditions you find, and ask: Why are these here? The answer to this question may lead back to the department supervisor—or it may lead to some other system weakness within the company—but the question should be explored fully and answered.

Consider, for example, that an unsafe ladder is discovered. The inspector should immediately ask the following: Why is this ladder here? Why was it not uncovered by our ladder inspection procedure? Why did the line supervisor allow it to remain here?

Answers to these kinds of questions begin to get at the true causes of accidents and fundamental system weaknesses. However, if very few inspections are occurring throughout industry, it appears that system weaknesses remain undetected.

■ Exhibit 6.1 **Typical inspection report**

INSPECTION REPORT

By John Jones, S.D. Date 1/16
cc: D. C. Anderson, J. C. Hansen

General conditions:

Housekeeping	Fair, some places need attention
Equipment	Generally good
Hand tools	Fair, some need repair
Lighting	Good
Ventilation	Good except in building
Floors	Good
Guards	See below

Recommendations:

Building 4, third floor	Change drill placement to avoid crowding. Guard power press No. 413.
Building 5, first floor	Change control switch on milling machine. Adjust tool nuts on all grinders so rest is at 1/8 inch from the wheel. Advise against hanging goggles on the machine for general use.

Recommendations and Reports

Instead of submitting lists of recommendations in reports (see Exhibit 6.1), safety professionals, managers, or supervisors might submit reports showing suggested changes in management procedures (Exhibit 6.2). The report shown in Exhibit 6.1 is still common today. It reflects no search for the causes of the unsafe conditions. The inspection is

■ Exhibit 6.2 **Suggested inspection report**

DEPARTMENT AUDIT

Department 43 Supervisor Bill Persons
Audit by John Jones, S.D. Date 1/16 cc: D. C. Anderson
J. C. Hansen

Appraisal of supervisor's inspection performance

Bill does an adequate job of inspecting the department, but doesn't seem to get to it often enough

Symptoms noted

	Discussed with supervisor?	Cause found?	Disposition
Conditions			
Missing press guard	Yes	Mainly training	Bill to handle
Defective ladder	Yes	SOP is weak	I will handle
Acts			
Operating without guard	Yes	SOP is weak	Bill will rewrite the SOP

Other

Suggestions Bill has agreed to a weekly scheduled inspection.

I will follow up in one month.

designed only to remove symptoms. The report in Exhibit 6.2 represents a better approach to the inspection.

The Primary Inspection Checklist

The line supervisor is often provided with a checklist to use for the primary inspection, such as that shown in Exhibit 6.3. This approach does not encourage an effort to trace back the symptoms to their true causes. This type of checklist should be reworded into a form that requires determination of some of the causes of the symptoms that have been unearthed (see Exhibit 6.4).

■ Exhibit 6.3 **Typical supervisor's inspection form**

SUPERVISOR'S INSPECTION FORM

Name________________________ Date__________________

Item:	Good	Poor	Disposition
Housekeeping Aisles			
Piling			
Floor surfaces			
Tools Condition			
Grounding			
Guards			
Personal protection			
Miscellaneous			
Ladders			
Slings			

■ Exhibit 6.4 **Suggested supervisor's inspection form**

SUPERVISOR'S INSPECTION FORM

Name____________________ Date____________________

Symptoms noted act / condition / problem	Causes why – what's wrong	Corrections made or suggested by you – by others

RECORDS

Accident records have always been and remain an important tool of the safety professional. Here again, ask this key question: *Why are we keeping records?* The answer to this question dictates the kind of records that should be kept. Much has been written over the years on accident record keeping, but the *reasons* for keeping the records are seldom identified.

Our discussion of records in this chapter is based on the premise that the primary answer to the above question is: We keep records in order to measure the supervisor's performance. Two distinct categories of accident records are generally kept in industry: accident investigation records and injury records.

Accident Investigation Records

The primary accident investigation function has always been assigned to the supervisor. Usually management provides the supervisor with a

■ Exhibit 6.5 **Typical accident investigation report**

SUPERVISOR'S REPORT OF INJURY

Name of injured ______________________________

Injury date ____________________ Time __________ A.M. – P.M.

Did injured return to work? __________ Time __________ A.M. – P.M.

Witnesses ______________________________
____

Nature of injury ______________________________
____

Where and how did the accident occur? ______________________________
____
____
____

Unsafe act or condition ______________________________
____

Measures taken in preventing a similar type of accident ______________________________
____
____
____
____
____

simple form on which to record the results of the investigation (see Exhibit 6.5).

Once again the supervisor is asked to investigate thoroughly and determine only *one cause* for the accident. This violates the multiple-causation principle. Asking the supervisor to identify only one act or condition ignores the principle that the act or condition is merely a *symptom* of the accident, not the cause.

■ Exhibit 6.6 **Suggested accident investigation report**

SUPERVISOR'S REPORT OF INJURY

Name of injured ______________________________

Injury date ______________________ Time __________ A.M. – P.M.

Did injured return to work? __________ Time __________ A.M. – P.M.

Witnesses ______________________________

____

Nature of injury ______________________________

____

Where and how did the accident occur? ______________________________

____

Identify:

Acts and conditions	Possible causes

Measures taken in preventing a similar type of accident
(List on the reverse side)

____

Supervisor's signature______________________ Department______________________

It is proper for line supervisors to investigate. But they should be allowed to determine what really happened, not be told to stop assessing an accident after identifying one contributory act or condition. The tools supervisors are given should encourage them to determine some of the *many* underlying causes. Perhaps the form shown in Exhibit 6.6 would better facilitate this type of investigation.

To tie the *measurement* or *accountability* idea into accident investigation, it seems logical that:

1. All accidents, not just lost-time accidents, should be investigated by the supervisor.
2. Management should receive the investigation form. (It must be transmitted up the line.)
3. At least five possible causes should be identified in each investigation.
4. At least two measures should be taken to prevent a recurrence.

Like the inspection tool, the primary investigation must be the responsibility of line management; however, in certain instances, it is desirable for staff safety people to carry out further investigations to determine causes. Usually more detailed digging can be done by the staff when it seems important that underlying causes be identified.

Some organizations routinely specify that the safety department will investigate all serious injuries, others leave that choice to the specialist. Some organizations will set up elaborate plans for detailed analyses of operational errors (some will do the same for potential errors).

DO WE GO FAR ENOUGH?

Actually, today's theories of accident causation suggest we should consider more in-depth examination of the interrelated causes of incidents; that is, that the incident is the result of: (1) a system's failure and (2) a human error.

System failure is concerned with most questions that traditional safety management might cover, such as:

- What is management's stated policy on safety?
- Who is designated as responsible and to what degree?
- Who holds the position of authority and what are they authorized to do?

- Who is held accountable and how is this determined?
- How are those responsible for safety selected?
- How are these safety supervisors measured for performance?
- What inspection systems are used to find out what went wrong?
- What systems are used to correct things found wrong?
- How are new people oriented?
- Is sufficient training given to employees?
- What are the standard operating procedures?
- What guidelines are used?
- How are hazards recognized?
- What medical program is in place?
- What records are kept and how are they used?

Human Error

The second and ever-present cause of an incident or accident is human error. Human error results from one misstep or a combination of three things: (1) overload, which is defined as a mismatch between a person's capacity and the load placed on him or her in a given situation; (2) a decision to err; and (3) traps that are left for the worker in the workplace.

The human being will not be able to avoid an accident if given a heavier workload than he or she has the capacity to handle. This overload can be physical, physiological, or psychological. To deal with overload as an accident cause, one must look at an individual's capacity, workload, and current situation.

In some situations, it seems logical to the worker to choose the unsafe act. Reasons for this might include:

1. Because of the worker's current motivational field, it makes more sense to operate unsafely than safely. Peer pressure, pressure to produce, and many other factors might make unsafe behavior seem preferable.
2. Because of the worker's mental condition, it serves him or her to have an accident.

3. The worker just does not believe he or she will have an accident (low perceived probability).

The third cause of human error involves traps that are left for the worker. This primarily involves common human factors. One trap is incompatibility; a worker errs because his or her work situation is not compatible with the worker's physique, or with what he or she is used to. A second trap is the design of the workplace—it is conducive to human error. Ultimately, a trap reflects the culture of the organization: what behaviors it encourages or discourages. Certain situations will be *error-provocative*.

MEASURING WHAT THE LINE SUPERVISOR DOES

First-line supervisory measures fit comfortably into most managerial systems. The following are some common safety approaches that use them.

Activity Measurement

Exhibit 6.7 lists some of the items that management might use to measure the line organization to determine what they are doing to prevent accidents from occurring. This is perhaps more important than the measurement of results because it measures line effort in controlling losses before accidents happen.

Management can measure line supervisors to see whether they are utilizing such techniques of accident control as toolbox meetings, job hazard analyses, inspections, accident investigations, incident reports, safety committees, and safety meetings. Management may require line supervisors to submit activity reports. When management measures these activities, it is setting up a system of accountability for activities. It is emphasizing to the supervisor that management wants performance in safety. It is, in terms of our model, raising the supervisor's perception of the probability that effort will be rewarded.

■ Exhibit 6.7 **Measurement of activities**

ACTIVITIES TO BE MEASURED

1. Safety meetings that supervisor holds
2. Toolbox meetings
3. Activity reports on safety
4. Inspection results
5. Accident investigations
6. Incident reports
7. Job hazard analysis

SYSTEMS TO USE

1. Regular reports
2. Sampling
3. SCRAPE
4. Performance rating

Exhibit 6.7 also lists some systems. One simple system would involve regular reports required from supervisors. An example of such a report is shown in Exhibit 6.8. Sampling is another possible system; it could involve statistical sampling or a simple sampling of acts and conditions made periodically by management using a form like the one shown in Exhibit 6.9

System of Counting and Rating Accident Prevention Effort (SCRAPE)

SCRAPE is a systematic method of measuring the effort made to prevent an accident. The SCRAPE rate indicates the amount of work

■ Exhibit 6.8 **SCRAPE activity report form**

	Points
Department ____________ Week of ____________	______
(1) Inspection made on ____________ # corrections ____________	______
(2) 5-minute safety talk on ____________ # present ____________	______
(3) # accidents ____________ # investigated ____________	______
Corrections ____________	
____________	______
(4) Individual contacts:	
Names ____________	
____________	
____________	
____________	______
(5) Management meeting attended on ____________	______
(4) New men (names): Oriented on (dates):	
____________ ____________	
____________ ____________	
____________ ____________	______

done by a supervisor and by the company to prevent accidents in a given period. Its purpose is to provide a tool for management that shows whether positive means are used regularly to control losses before accidents occur. The performance rating is similar in that it attempts to quantify the supervisor's effort (as well as the supervisor's role perception and ability level).

The first step in SCRAPE is to determine what specific functions the company wants the line manager to perform in safety. Normally

■ Exhibit 6.9 **Report of supervisor's safety activities**

REPORT OF SUPERVISOR'S SAFETY ACTIVITIES

Supervisor ______________________ Department ______________

Date ________________ This report covers ________ to __________

Inspections Made

Inspection date ___________ No. hazards corrected _____ No recs. to mgmt _____

Inspection date ___________ No. hazards corrected _____ No recs. to mgmt _____

Inspection date ___________ No. hazards corrected _____ No recs. to mgmt _____

Comments:

Meetings Held

Toolbox meetings Date ___________ No. employees _____ Subject ________

Date ___________ No. employees _____ Subject ________

Date ___________ No. employees _____ Subject ________

Other meetings (explain):

Accidents Investigated

Number of accidents investigated this period ______________________

Number of hazards corrected ______________________

Number of recommendations to management ______________________

Comments:

Employee Contacts

New employee safety orientation

Name ____________ Date _______ Name ____________ Date _______

Name ____________ Date _______ Name ____________ Date _______

Name ____________ Date _______ Name ____________ Date _______

Other employees

Name	Date	Subject	Name	Date	Subject

Use Safety Materials

List materials used this period

Accident record	This period	Year to date
Number first-aid cases		
Number doctor cases		
Number lost-time cases		
Man-hours worked		
Frequency rate		
Severity rate		

Comments:

this falls into six categories: (1) making physical inspections of the department, (2) training or coaching the people, (3) investigating accidents, (4) attending management meetings, (5) establishing safety contacts with the workers, and (6) orienting new people.

SCRAPE allows management to select which of these functions it wants supervisors to perform and then to determine their relative importance by assigning values to each:

Item	Points
Departmental inspections	25
Training or coaching (e.g., 5-minute safety talks)	25
Accident investigations	20
Individual contacts	20
Meetings	5
Orientation	5
Total	100

Depending on management's desires, the point values can be increased or decreased for each item. They should, however, total 100 points.

Every week, each supervisor will fill out a small form (see Exhibit 6.8), indicating weekly activity. On the basis of this form, management spot checks the quality of the work done in all six areas and rates the accident-prevention effort by assigning points between 0 and the maximum point value for that item.

SCRAPE can provide management with information on how the company is performing in accident prevention. It measures safety activity, not a lack of safety. It measures activity before the accident, not after. Most important, it makes management define what safety performance it wants from supervisors and measures to see that it is achieving what it wants. SCRAPE is a system of activity accountability.

MENU

In the MENU approach, supervisors or managers are required to perform some mandatory activities (inspections, contacts, etc.) and are also allowed to select other activities they are comfortable with. They are measured and rewarded for the achievement of both.

Safety by Objectives

Safety by objectives (SBO) is one of the most effective of the approaches adopted by organizations. It is based on the old management-by-objectives (MBO) approach. The concept is simple: managers devise specific objectives initially (results, activities, or both), then with their immediate boss they reach a mutual agreement on those objectives. Thus, under SBO the measurement problem is simple; it is a part of the whole process. What is measured is simply whether the manager has in fact met the objectives. Only enough record keeping is needed to keep track of progress toward the agreed-upon goal (see Chapter 7).

Among organizations, the MENU approach is the most popular. SCRAPE seems to be selected by more hierarchical, "top-down" kinds of companies. SBO is typically selected more by participative, "bottom-up" organizations. Since most organizations are somewhere in between these two, they tend to select the MENU approach. (See Appendix D, Case Study 5.)

RESULTS MEASURES

Results measures can be used either before an accident occurs or afterward. After-the-accident measures might be considered measures of failure.

With failure measures, we can count incidents, accidents, or injuries, and we can count these at various levels of severity. For example, we can count only fatalities, only lost-time injuries at the level of seven or more days, only injuries resulting in one or more days lost, only those cases requiring the attention of a doctor, or perhaps only cases requiring first aid. We can even start counting close calls.

What we choose to count will change our results markedly. For lower-level management, it is best to use a measure that gives us lots of numbers. For a supervisor of ten people, a measure of close calls as well as injuries makes more statistical sense and is more meaningful than one that counts only fatalities or only lost-time injuries, but here we face the problem of input integrity (how do we get all the near misses?).

Before-the-Fact Measures

Before-the-fact measures rate the results of supervisor action before an accident occurs. Our earlier example of a periodic inspection of a supervisor's work area to ascertain its physical status produced a list of any unsafe conditions, telling us how many things are wrong. By measuring the employees' work behavior, how well a supervisor gets through to the people in the department can also be determined (see Exhibit 6.10). Safety sampling is used for such measurements.

Results measures can:

- be constructed to give swift feedback
- attract attention
- be sensitive to change
- provide recognition of good performance.

However, we must use care in selecting results measurements. Since we are judging performance by some means other than the performance itself, we must be sure that what we are looking at is closely related to performance. With failure measures, this close relationship is quickly lost at lower levels of the organization. Before-the-fact measures tend to retain the close relationship much better.

SAFETY SAMPLING

One of the best methods of fixing accountability, using statistical methods, is safety sampling. It measures the effectiveness of the line manager's safety activities, but not in terms of accidents. It measures

■ Exhibit 6.10 **Manager's safety survey**

MANAGER'S SAFETY SURVEY

PART I. UNSAFE ACT SAMPLING

A. Number of Safe Observations

B. Number of Unsafe Observations

Unsafe act	Dept.				
No safety glasses/unauthorized glasses					
Not using machine properly					
Machine unguarded/guard not adjusted					
Not using tools, jigs, pushsticks, etc.					
Working near tripping hazard					
Improper use of air nozzles					
Improper use of hand tools					
Loose clothing near machine					
Improper lifting/positioning					
Climbing on racks					
Unsafe loading/piling/storing					
Using defective equipment					
Other (specify)					

C. Percentage Last period Last year

Dept.	%	Dept.	%	Dept.	%

PART II. UNSAFE CONDITION INSPECTION

D. Violations

	Dept.				
Machine guards					
• Transmission					
• Point of operation					
o Missing					
o Not properly adjusted					
Electrical					
• Cords					
• Grounds					
Flammables					
• Amount					
• Use					
• Extinguishers					
• Exits					
Falls					
Floor surfaces, objects					
Air nozzles					
Hand tools					
• Guards					
• Grounds					
Others (specify)					

E. Number Last period Last year

Dept.	No.	Dept.	No.	Dept.	No.

effectiveness before the fact of the accident by taking a periodic reading of how safely the employees are working.

Like all good accountability systems or measurement tools, safety sampling is also an excellent motivational tool, since line supervisors find that it is important for their employees to be working as safely as possible when the sample is taken. To accomplish this, they must carry out some safety activities, such as training, supervising, inspecting, and disciplining. Many organizations that have utilized safety sampling report a substantial improvement in their safety record as a result of the increased interest in safety on the part of line supervisors.

Safety sampling is based on the quality-control principle of random-sampling inspection, which is widely used by inspection departments to determine the quality of production output without inspecting every piece. For many years, industry has used this technique in which a random sampling of a number of objects is carefully inspected to determine the probable quality of the entire production line. The degree of accuracy desired dictates the number of random items that will be carefully inspected. The greater the number of items inspected, the greater the accuracy.

Safety sampling is a tool that has the potential of providing management with information about which line supervisors are doing their jobs in safety and which are not. It is a method of systematically observing workers in order to determine what unsafe acts are being committed and how often they are occurring. The results of these observations are then used to measure the effectiveness of line safety activities. This type of sampling has been used even by the operators of theme park rides (Lyon, 2001).

In utilizing the tool of safety sampling, first a list of unsafe codes is prepared. The most common unsafe acts are listed on a form, such as the one shown in Exhibit 6.11. Next, a sample is taken by walking rapidly through the operation and observing each employee quickly. An immediate decision is made about whether each employee is working safely. If the employee is working safely, it is checked off on a theater counter as one safe observation. If the employee is working unsafely, the observer will record this as one unsafe act. The third step is to validate the sample statistically to determine whether there are enough observations to constitute a representative sample. The fourth and final step is to prepare a report for management. It shows each supervisor's rating, expressed as a ratio of safe to unsafe acts. This can be compared with past records and with the records of other departments. Management then can judge line performance and apply whatever action is necessary.

Procedure

1. **Prepare a Code.** This list of unsafe practices is the key to safety sampling and supervisor training. It contains specific unsafe acts that occur in your plant. These are the "accidents about to happen." This list is developed from the accident record of each plant. In addition, possible causes are also listed, and the code is entered on an observation form.

2. **Take the Sample.** With the code attached to a clipboard and using a theater counter, start the sample. The inspector identifies the department and the supervisor responsible. He or she then proceeds through the area, observing each employee who is engaged in some form of activity. The inspector instantaneously records a safe or an unsafe observation of the employee.

   Each employee is observed only long enough to make a determination, and once the observation is recorded, it should not be changed. If the observation of the employee indicates safe performance of the job, it is counted on the theater counter. If the employee is observed performing an unsafe practice, a check is made in the column, indicating the type of unsafe practice by the element's code number.

3. **Validate the Sample.** The number of observations required to validate a sample is based on a preliminary survey and the degree of desired accuracy. The following data must be recorded

■ Exhibit 6.11 **Example of a list of unsafe acts**

Page 1 of 2 — Department

SAMPLING WORKSHEET

Safe Observations

Unsafe Acts

		DC & service	Maint. power	Tool room	Foundry & pattern	Stock & shipping	Rotor	Shaft	Punch press	Body & frame	Bracket	Small winding	Large winding	Small assembly	Lg. assembly & pck.
1	Improper lifting														
2	Carrying heavy load														
3	Incorrect gripping														
4	Lifting without protective wear														
5	Reaching to lift														
6	Lifting and turning														
7	Lifting and bending														
8	Improper grinding														
9	Improper pouring														
10	Swinging tool toward body														
11	Improper eye protection														
12	Improper foot wear														
13	Loose clothing – moving parts														
14	No hair net or cap														
15	Wearing rings														
16	Finger/hands under dies														
17	Operating equip. at unsafe speeds														
18	Foot pedal unguarded														
19	Failure to use guard														
20	Guard adjusted improperly														
21	Climbing on machines														
22	Reaching into machines														
23	Standing in front of machine														
24	Leaning on running machine														
25	Not using push sticks (jigs)														
26	Failure to use hand tools														
27	Walking under load														
28	Leaning – suspended load														
29	Improper use of compressed air														
30	Carrying by lead wires														

■ Exhibit 6.11 (Cont.)

Page 2 of 2		Department													
SAMPLING WORKSHEET Safe Observations Unsafe Acts		DC & service	Maint. power	Tool room	Foundry & pattern	Stock & shipping	Rotor	Shaft	Punch press	Body & frame	Bracket	Small winding	Large winding	Small assembly	Lg. assembly & pck.
31	Table too crowded														
32	Hands and fingers between metal boxes														
33	Underground power tools														
34	Grinding on tool rest														
35	Careless alum. splash														
36	One bracket in shaft piling														
37	Feet under carts or loads														
38	Pushing carts improperly														
39	Pulling carts improperly														
40	Hands or feet outside lift truck														
41	Loose material under foot														
42	Improper piling of material														
43	Unsafe loading of trucks														
44	Unsafe loading of skids														
45	Unsafe loading of racks														
46	Unsafe loading of conveyers														
47	Using defective equipment														
48	Using defective tools														
49	Evidence of horseplay														
50	Running in area														
51	Repair moving machines														
52	No lock-out on machine														
	TOTAL UNSAFE ACTS:														
	ADDITIONAL UNSAFE ACTS:														
53															
54															
55															
56															
57															

on the preliminary survey: total observations and unsafe observations. The percentage of unsafe observations is then calculated. Using this percentage (P) and the desired accuracy (Y), which we will determine as a + or – 10%, we can calculate the number of observations (N) required by using the following formula:

$$N = \frac{4(1 - P)}{Y^2(P)}$$

Where N = total number of observations required
P = percentage of unsafe observations
Y = desired accuracy.

For example, if the preliminary survey produced the following results: (1) total observations equal 126 and (2) unsafe operations equal 32, the percentage of unsafe to total observations would be 32 divided by 126, which is 0.254 or 25%. Thus

$$N = \frac{4(1 - P)}{Y^2(P)}$$

$$N = \frac{4(1 - 0.25)}{(0.10)^2(0.25)}$$

$$N = \frac{3}{0.0025}$$

N = 1,200 (number of observations required)

To give effective results, this study must have a minimum of 1,200 observations.

4. **Report to Management.** The results can be presented in many different forms. However, the report should include the following:

   a. Total percentage of unsafe activities by department and by shift.
   b. Percentage of unsafe activities by supervisor, general foreman, or superintendent.
   c. Number and type of unsafe practices observed.
   d. Types and number of unsafe observations which are supervisory responsibilities.

   For examples of these results, see Exhibits 6.12 and 6.13.

■ Exhibit 6.12 **Example of safe practices sampling report**

SAFE PRACTICES SAMPLING REPORT

Plant I　　　　Period covered October

Department supervisor	Unsafe practice code number													Observations		%
	1	2	7	9	11	17	26	34	36	59				Total	Unsafe	unsafe
E. Jones–Supt.														1.094	39	3.4
Smith–gen. for														246	9	3.5
Jolas			1					1	1					90	3	3.2
Johnson		3			1			1	1					156	6	3.8
G. Mairathur														226	11	4.6
Mantle				1		1		1	1					101	4	3.8
Williams				1	1									53	2	3.6
Nedstrom					1			1	1	2				72	5	6.5
Mack														284	13	4.4
Peters		1			3	1								96	5	5.0
Sadieri							3	1	1					73	5	6.4
Albert	1		1											64	2	3.0
Anderson	1													51	1	1.9

■ Exhibit 6.13 **Example of safety sampling report**

SAFETY SAMPLING REPORT

Plant I Month of October

Department	Total observations	Unsafe observations	% of unsafe activity	
			This month	Previous month
Manufacturing–Prod.	442	77	17.4	12.1
Press	1,815	244	15.3	19.7
Assembly	1,699	59	4.0	4.0
Welding	322	70	21.0	11.2
Subtotals	4,276	450	14.4	14.2
Production Eng.	339	55	16.2	21.5
Plant Engineering	341	51	14.9	26.7
Subtotals	680	106	15.6	23.4
Plant Totals				

STATISTICAL CONTROL TECHNIQUES

As mentioned previously, the accident rate may fluctuate from period to period and still reflect nothing more than chance variation. There are times, however, when a fluctuation occurs because something is different in the system. This could occur during a period of high turnover, resulting in a greater than usual number of untrained workers, or during a period of sudden increase in production pressure, resulting in a higher than normal pressure to perform.

Safety professionals need a tool that will enable them to detect the presence of new accident causes. They should be relatively unconcerned with minor fluctuations in the accident picture when they are sure that the situation has remained stable. However, they need something that will alert them quickly when the situation has become unstable.

Statistical control techniques can perform this essential job. They will signal a significant change in the accident process, giving the safety professional assurance that a change has taken place. The safety specialist can then identify causes (see the Appendix C).

Safety experts have proposed a number of both activity and results indicators, or a combination of these indicators, which may be worth considering. Here are a few.

From Kirk Rowell, the Safety Ratio:

> The following ratio, when used with a safety incentive program, is a simple and useful tool to reward employees for reducing accidents while, at the same time, encouraging the reporting of nonlost-time injuries and the execution of positive safety measures such as incident reporting:
>
> $$\frac{\text{Total \# incidents (excluding lost time)}}{\text{\# lost-time incidents}} = \text{Safety Ratio}$$
>
> In this equation, the *total number of incidents* refers to any safety-related reporting that a company may perform. Examples include the number of hazard reports, near-miss reports, safety suggestions, and incidents that require first aid, medical treatment, or restricted job duties.
>
> *Lost-time incidents* means any incident that causes an employee to miss work.
>
> There are only two ways to increase the safety ratio: increase the reporting of nonlost-time incidents (numerator) or decrease the number of lost-time incidents (denominator).

From Jack Toellner:

Many safety professionals spend a significant amount of time gathering, analyzing, and reporting statistics. If these efforts do not directly lead to improved performance, a site's safety resources are not being maximized.

The safety profession's goal should be to improve both short- and long-term performance. Fewer accidents lead to fewer injuries and illnesses, which means fewer lives disrupted. As frustrating as they may be at times, trailing indicators offer some insight into safety performance. Used correctly, resulting data can help management and workers better understand overall performance trends and the significance of relatively minor events.

Leading indicators are used to focus resources on preventive actions. They:

- allow management to actively demonstrate commitment and leadership;
- enable workers to get involved with measurable processes; and
- focus resources on accident prevention processes.

From Amoco:

Success in any endeavor is aided by having a clear target or objective in mind. In managing safety, objectives and performance criteria are needed to allow assessment of our efforts to improve workplace conditions and eliminate injuries. These objectives are also used to establish accountability and to drive the continuous improvement cycle. This element of a safety management system must describe how objectives and performance criteria will be selected, how appropriate levels or targets will be established, and how the performance will be tracked, assessed and reported.

References

Amoco, "Template for a Safety Management System," Amoco, undated.

Lyon, V. "Behavior Sampling of Theme Park Ride Operators," *Professional Safety*, October 2001.

Petersen, D. "Human Error, Safety's Next Frontier," *Professional Safety*, December 2003.

Rowell, K. "The Safety Ratio," *Occupational Hazards*, vol. 62, no. 6 (June 2000).

Toellner, J. "Improving Safety & Health Performance: Identifying and Measuring Leading Indicators," *Professional Safety*, September 2001.

CHAPTER 7

The Key Micro Measure: Performance to Goal

THE KEY DRIVER AT LOWER LEVELS of an organization is performance to goal: Did the employee meet the objectives set?

There are several categories in which to set goals:

1. **Routine Goals.** These are the objectives that will relate to the regular, repetitive, commonplace activities which are the normal, basic duties and requirements of the job. For instance, a routine goal for department supervisors might relate to the number of accidents that occur to their people or the number of inspections they will make.

2. **Project Goals.** These refer to new or special projects that might be undertaken by a manager in a coming time period. They also could relate to the problem-solving or emergency-action kind of activities that are a part of the supervisory job.

3. **Creative Goals.** While less commonly used, these could refer to the kind of objectives for innovation or improvements in the department, in the management system, or at another level of the company.

4. **Personal Goals.** Also less common, but sometimes used, are goals that relate to personal efforts that might be made to supplement job skills with new skills to increase personal effectiveness.

All of these are certainly feasible areas for goal-setting in safety management. In most cases, goals will be set in more than one of the above categories. Perhaps a number of goals will be set in the routine area of a supervisor's job, but often objectives will be set in the other three categories as well.

CRITERIA FOR OBJECTIVES

If this objective-setting process is to work in an organization, the goals that are set should attempt to produce some value for the company. In this section, we concentrate on the setting of objectives and try to identify what constitutes a "good" objective as opposed to a "poor" objective, and then determine what criteria we might be able to establish for a good objective. First, we can set criteria aimed at how the objective is devised and written, and then we can set some further criteria aimed at the relationship between the two participants (the subordinate and superior managers) in the process.

A good objective will address four criteria pertaining to its direction, its individuality, its measurability, and its realism in the situation that exists:

1. **Zeroing of objectives.** This refers to the fact that a good objective aims at a particular and specific area of performance, as opposed to rather general, or "shotgun," objectives that aim at broad, nonspecific performance areas. For instance, an objective like "to be a better supervisor next month" is so general that it is useless.

2. **Individuality of objectives.** This refers to the specificity of the objective to the subordinate for whom it is set. An individual objective is one that requires results and performances that the individual subordinate can do alone. This means that the subordinate must have enough control over those things that need to be done to attain the objective. An objective for the staff safety director to improve the record in department A

is an objective that is not individual to the safety director because, to attain the objective, that director is dependent on the performance of the supervisor of department A.

3. **Measurability of objectives.** If we are going to assess the performance of a manager toward reaching a goal, it must be measurable. For example, a goal "to become a better listener in the next six months" is immeasurable and therefore a poor objective.

4. **Realism of objectives.** Objectives must be based on facts and on analysis of past performances, or on some other valid data, so they are not merely dreams and wishes. To be effective, objectives must be attainable. The personal goal "to be a millionaire next year" is simply not a realistic objective.

All of the above criteria speak to the objective itself. Other standards may be established to decide whether the objective has been reached. In this area, we might say that a good objective would meet three additional criteria:

1. It would be leveled; that is, it should be aimed directly at the organizational level of the person. A goal that speaks to the results of a supervisor's department in a specific area is at the supervisor's level if he or she is the direct supervisor of that department. Requiring a supervisor to "help the corporation attain the desired frequency rate" is not an objective that speaks to a supervisor's level. An objective at the supervisory level would spell out specific actions that he or she can carry out that will help attain that goal.

2. It must be understood both by the superior and the subordinate. If it is not clearly grasped by both sides, it is a poor objective. Misunderstandings are one of the biggest reasons for unattained objectives.

3. It must be mutually agreed upon by both parties. Neither party to the process may impose his or her will on the other. Without mutual agreement, the objective is relatively useless.

As an example, review the objectives in Exhibit 7.1. It should be easy to determine whether each meets or fails to meet the above objective criteria.

■ Exhibit 7.1 **Sample of possible objectives**

	Criteria						
Objective	Zeroed	Individual	Measurable	Realistic	Leveled	Understood	Mutually accepted
To respond faster and more accurately to work orders coming into the department.	–	–	–	–	–	–	–
To reduce rejections from quality control due to faulty assembly in my department by 5 percent over the next three months.	–	–	X	X	–	X	–
To increase sales of product B in my territory by 37 percent over last year, within three months, and without a reduction in the sales volume of my other products.	X	X	X	X	X	X	X
To pull together so that the number of customers we wait on in our department can be increased by 5 percent.	–	–	X	X	–	X	X
To respond to 96 percent of the work orders coming into the department within four hours of the arrival in the department, with no order taking more than eight hours to process, and with an error rate of less than 5 percent.	X	X	X	X	X	X	X
To improve your acceptance by the line managers as an advisor on safety problems as evidenced by an increasing number of people who know your name when I bring it up, who have favorable things to say about your advice, and who quote you in their discussions with me.	X	X	X	X	X	X	X

WHY OBJECTIVES ARE NOT MET

Obviously, the installation of an objective-setting approach does not solve all the problems that an organization has relating to performance of line managers in safety. Often, under such an approach, objectives are set but not reached. In that case, an analysis is in order to determine what went wrong. The following are some possible reasons why the objectives were unattainable:

1. The objective was poorly constructed. Most likely, it did not meet the previously defined criteria. In many cases, this is the reason an objective has not been reached.

2. The employee striving to meet the objective had insufficient skills to meet the goal. Perhaps the person simply did not have the necessary ability and knowledge to carry out the objective. This often happens, even though both parties did not realize it during the objective-setting process.

3. There may have been a change on the part of the subordinate following the agreement on the objective. For some reason, the subordinate lost belief in or dedication to the original objective.

4. There may have been some barrier to performance in the line organization or at the executive level (e.g., a shift in emphasis, a reemphasis on other items, lack of necessary resources to attain the goal, etc.).

This key measurement of performance to goal is useful at almost any organizational level, but is emphasized at the supervisory level because there it most directly drives performance. For a more detailed discussion of measurement criteria for line supervisors, see Chapter 5.

REFERENCES

Petersen, D. *Safety by Objectives, What Gets Measured and Rewarded Gets Done.* New York: Van Nostrand Reinhold, 1996.

CHAPTER 8

Macro Measures: Grading Overall System Improvement

As we ascend the organizational ladder, our criteria for measures of safety performance change slightly, and the measures change also. Since the database grows with each step up, we can begin to use results measures with more confidence. At the lower levels we tend to use performance measures to a large degree to ensure that we are not rewarding merely luck. At the top levels we become much more results-oriented and tend to look less at performance. At the middle level we look at both performance and results.

As we use it here, the term "middle manager" is a bit of a catchall, alluding to an employee at almost any level of management who is *over* the first-line supervisor and *under* the top executive. Thus, a middle manager could be a line manager over two or three supervisors, a plant superintendent with ten supervisors, or a manufacturing manager or vice president to whom several levels report. Middle managers do not have ultimate policy decision-making authority; rather, they carry out the plans, policies, and procedures decided upon above their level. Similarly, they are not responsible for daily supervision of employees; people under them do that.

As top managers judge middle managers, they want to make sure that the middle managers maintain a strong safety emphasis, so they use activity measures, but they probably also judge them heavily by the results they have achieved.

We use different measures as we go higher up the organizational ladder because results statistics tend to become more reliable and valid at the upper levels. At the lower levels the database is simply too small to provide statistical significance, and at the supervisory level we want to motivate and reward performance, so we must measure activities whenever possible.

MEASURING SYSTEM IMPROVEMENT

Basically, three methods are used to measure safety in any organization:

1. The accident record (results measures that are subsets of the injury records, and which are discussed in Chapter 9)
2. Scored assessments (audits)
3. Perception surveys.

Audits

Scored assessments are tools used to evaluate the effectiveness of a safety management system. They have the advantage of providing a numerical basis for the measurement of efforts, but they can be a prescriptive approach to safety management that is sometimes inflexible. In addition, some managers question whether they are measuring the right items, and note that achieving a score becomes more important than achieving results. The audit scores must be correlated to the injury record to be really predictive, otherwise the audit may be measuring things that are nonproductive.

The audit concept has been around for a long time. In the 1950s, we began thinking about the audit—the predetermination of what must be done, of who should do what, and how we measure these predetermined actions. This, of course, starts with the determination of the contents of the safety system.

As the audit concept became widespread, the idea of the packaged audit gained popularity. Perhaps starting in Canada with the book *Total Loss Control* by Jack Fletcher, the concept evolved throughout the world. Examples of the packaged audit are the Nosa International System in South Africa, the British Safety Council System, the Integrated Loss Control, Inc. (ILCI) Five Star Program, the International Standards Organization (ISO) programs, and others. The idea behind the packaged audit is that there are certain defined things that must be included in a safety system to get a high rating or the greatest number of stars.

How does this thinking jibe with the research? Not too well.

- A NIOSH study in 1978 identified seven crucial areas needed for safety performance (Diekemper and Spartz, 1979). Most are not included in the packaged programs.
- Foster Rinefort's doctoral dissertation at Texas A&M suggested there was no one right set of elements. (Rinefort, 1974)
- National Safety Council studies in 1967 and 1992 suggested many of the "required" elements of packaged programs were quite suspect in terms of effectiveness. (Planek, 1993)
- A major study recently done by the Association of American Railroads conclusively showed that the elements in most packaged programs had no correlation with bottom-line results. (Bailey, 1988)

The research questions the validity of the concept of packaged safety audits. It does not question the value of the audit concept itself.

Some other results were truly surprising:

- One Canadian oil-company location (Imperial Oil) consciously chose to lower its audit score and then found that its frequency rate significantly improved.
- One chain of U.S. department stores (Venture Stores) found no correlation between the audit score and workers' compensation losses. They found a negative correlation between the audit score and public liability loss payout.

Apparently, there had been little effort to correlate the packaged audit results with the accident record; safety people were buying into an unproven concept.

THE SELF-BUILT AUDIT

The self-built audit could well be the answer to how to construct an audit that accurately measures the performance of a safety system. (Examples are included in Appendix A.)

The process of developing an audit consists of:

1. Defining the safety-system elements
2. Defining the relative importance of each element (weighting)
3. Defining the questions to find out what is happening.

Nevertheless, some painful lessons can be learned from exercising the above process. In one organization, an audit that looked great in the development process ended up with no correlation to results.

The audit *must* be tested against reality. In one organization, the accountability section was heavily weighted, but not enough attention was paid to the quality-of-performance section. The result was a paper program only.

Esso Resources Canada Ltd. (Imperial Oil) is one of many organizations that has moved away from the packaged outside audit and to the internally constructed audit, which places major emphasis on employee interviews rather than checking paperwork.

Creating an audit that works well for the organization is a critical first step, but there are many suitable ways to go about implementing the audit:

- by sending in an individual or a team
- by using an insider or an outsider
- by using a staff person or a line manager
- by announcing the audit or by coming in unannounced
- by quantifying results or by taking a verbal approach
- by checking on a department or by using a coaching approach
- by examining only or by examining and questioning
- by looking at records and reports or by interviewing employees
- by using questionnaires or by not using questionnaires
- by choosing what to look at and what not to look at
- by determining whom to talk to and whom not to talk to.

Each choice makes a major difference, and each is a decision that must be made based upon climate and culture, beliefs, and the climate desired for the future.

Peggy Farabaugh describes a study of occupational health and safety (OHS) systems made at Tulane University:

> During the past several years, OHS managers and their management have been devoting increased attention to the development of occupational health and safety management systems (OHSMS). There are three major reasons why. First, the International Organization for Standards (ISO) has established a quality assurance system model (ISO 9001) and an environmental management system model (ISO 14001) that have been widely accepted around the globe. Second, OSHA has developed a Voluntary Protection Program and related programs that are based on the agency's Safety and Health Voluntary Program Management Guidelines. Finally, successful organizations have reported improvements in workplace health and safety, along with significant savings from the implementation of OHSMSs.
>
> One student population of OHS managers from diverse areas of industry, agriculture, government, R&D, and academia was perfect for providing an excellent cross-sectional view of what American businesses are doing in terms of implementing systems for occupational health and safety management.
>
> OHSM Models
>
> The OHSM models we had studied included OSHA's existing and proposed programs, the American Industrial Hygiene Association OHSMS guidance document, and the Chemical Manufacturers Association's Responsible Care® program, among others. The scope ranged from a narrow focus of OSHA compliance to integration of the safety function within the entire organization.
>
> The major difference noted among the models included scope, involvement of third-party audits, and accessibility.
>
> - Scope. Does the model focus strictly on environment, safety, and health, or is it designed to integrate ES&H into the broader construct of overall management systems?
> - Third-party audit practices. There is considerable controversy in the OHSM world these days surrounding the issue of third-party

> audits. Is this a method to provide unbiased judgment of the merits of a program, or does it primarily provide employment for the consultants who will be hired to make those judgments? (Farabaugh, 2000)

Perception Surveys

Measurements should indicate the level of safety-management-system acceptance among line employees, determining whether they believe in the system and have a real voice in its implementation. Perception surveys assess underlying attitudes and observations that influence acceptance of a safety management system, and as a result, determine the system's overall effectiveness. They offer the following advantages:

- They provide effective evaluation of the subjective factors that influence the safety management system: attitudes, value perceptions, and the effectiveness of specific systems.
- They can determine where significant gaps exist between management and employee perceptions.
- They can pinpoint problem areas that might otherwise be overlooked.
- Comparison of baseline and future surveys can determine the effectiveness of new safety initiatives.

However, there are also some disadvantages:

- They do not determine compliance with regulatory standards.
- Third-party assistance could be necessary to properly evaluate and utilize survey results.

The concept behind a perception survey is to ask people in the organization, anonymously, what they think. This can be done using a statistically valid survey tool and/or the interview method.

Top management frequently operates on the illusion that they are making some headway and that things are not as bad as they really seem. This can present a significant obstacle to the change process, which is comprised of three main steps:

1. Assessing your current reality (the current safety culture)
2. Identifying necessary changes
3. Planning and implementing specific changes.

The perception survey makes it possible to ask the workers what they see and how they feel in a comprehensive and complete way. It may be the first time they feel they can be really truthful. Basically, it provides a pipeline of information directly from the workers to upper management, bypassing the blockage of middle management. After the information is transmitted, the change process begins, and it must involve people from all levels in the organization to avoid resistance.

This involvement is the key to achieving change. Creating a new safety culture is a continuous process that never really ends. All of the issues identified by the perception survey cannot be solved at once. The approach is to form problem-solving work groups representing a cross-section of the organization. Work groups made up of managers, supervisors, and employees are formed around each of the lowest-rated safety categories to analyze the problems and devise a plan that resolves them.

The hardest part of the entire process is deciding to do the perception survey. Once completed, the results of the survey provide top management with the ammunition that will launch the remainder of the change process. Even those most set in their ways have trouble rejecting real data.

Data accumulated through these surveys can be descriptive of how truly effective safety systems are. If a pattern exists in the surveys of many organizations over time, it might suggest that overall approaches to safety need to be reassessed and, perhaps, altered.

A fundamental difference exists between what perception surveys tell an organization and what research or benchmarking efforts reveal.

Most safety research asks safety professionals or managers what they think works or does not work. Similarly, in the search for "best practices," benchmarking efforts also tend to ask safety professionals and managers what they feel works or does not work.

Conversely, perception surveys query hourly employees. The successful use of perception-survey data suggests that the key reality is hourly employee perception. Supervisor and/or manager perception is measured only to determine how far from reality it is; this exercise is useful within an individual organization because it reveals just how removed management is from shop-floor employees.

There are a number of perception surveys available. Exhibit 8.1 is a partial example of a survey developed with federal funding, thus making the questions available to anyone. It consists of 100 yes-or-no questions that measure 21 categories. In addition, these 21 categories can measure the nine components. This survey is relatively inexpensive and can be administered and scored internally by an organization once the software has been purchased, or it can be scored and interpreted by an outside evaluator.

■ Exhibit 8.1 **Sample perception survey**

The Perception Survey

21 Safety Categories

This sample questionnaire may be useful to anonymously assess worker's perceptions of their workplace's safety culture. Once employers have real data, work groups composed from every echelon of the organization can work to improve the areas that may need it.

9 Elements

1. Management's credibility, including:

Support for safety	Is the whole organization seen as working together to create a safe work environment?
Perception of culture	Has a climate been created that is conducive to adopting safe attitudes and work habits?
Goal setting	Do workers and management meet together to formulate behavior-oriented safety goals?
Operational procedures	Are safe procedures seen as both necessary and adequate by all levels of the organization?
Employee perception of credibility	Is management seen as wanting safe performance?

■ Exhibit 8.1 **(cont.)**

2. Supervisory performance, including:	
Quality of supervision	Are supervisors perceived to be competent in accident prevention?
Supervisory training	Are supervisors perceived as well trained and able to handle problems related to safety?
Employee recognition	Is good safety performance recognized?
Discipline (coaching)	Is the company perceived as taking a fair approach to handling rules and infractions?
Safety contacts	Are there regular safety contacts with all employees?
Inspections	Are there regular inspections of all operations?
3. Employee involvement and participation	
Involvement of employees	Are there opportunities for employees to become involved in safety through such means as quality improvement teams, ad hoc committees, or effective supervision?
4. Employee training, including:	
New employee orientation	
Employee training	Do employees feel that they receive adequate training in how to work safely?
New employees	Are new employees thoroughly trained in safety?
5 Employee attitudes and behavior	
Attitudes toward safety	Is there a positive attitude toward safety at all levels of the organization?
6. Communications, including:	
Awareness programs	
Communication	Do managers and employees communicate freely on safety issues?
Safety	Do you hold awareness programs that stress safety both on and off the job?
7. Accident investigation procedures	
Accident investigation	Does your safety system deal positively with the investigation of accidents?
8. Hazard control	
Hazard correction	Is there an effective system for dealing with reported hazards?
9. Stress	
Stress	Does the organization have an abundance of stress claims?
Alcohol and drug abuse	Are employees with substance-abuse problems allowed in the workplace?

INTERPRETING PERCEPTION SURVEYS

In recent years, many organizations have performed perception surveys provided by various sources. While it is not possible to accumulate data from different surveys, data from companies using the same survey can be discussed.

Although norms cannot be established based on overall scores from surveys conducted within an array of organizations (including furniture manufacturers, railroads, chemical and petrochemical facilities, paper mills, food processing plants, and construction projects), the scores can be interpreted and shared with each organization.

Scores are communicated as "% positive" for each category. A survey can measure by unit, location, or craft within each company to reveal any similarities or differences. By-category maximum, minimum, and mean scores are calculated for a number of companies in order to provide a picture against which an organization can compare itself.

The remainder of this section discusses survey data from 160 companies, which cumulatively employed millions of people at all levels. These results depict how these firms have succeeded or failed in their safety efforts, as judged by the people who truly count: the hourly employees.

Survey Results

A score below 70% positive in a category at the hourly employee level suggests the need to examine the organization's safety activities. Such a score indicates that three of ten employees do not believe the system is working well. A score below 60% positive is a red flag, indicating that the system needs help.

Analysis of this data reveals some similarities in safety-system-element effectiveness. For example, as a rule, firms are not highly successful in categories that score, on average, below 60% positive for hourly employees only. The same can be said for categories that score below 70% positive:

- Recognition – 61.9% positive
- Discipline – 61.5% positive
- Inspections – 62.3%

Borderline categories include:

- Supervisory training – 71.4%
- Motivational programs – 72.4%
- Operating procedures – 72.5%
- Substance abuse – 72.5%

Results for other categories include:

- Employee involvement – 74.6%
- Support for safety – 74.6%
- Climate – 76.6%
- Quality of supervision – 76.6%
- Attitudes toward safety – 76.6%
- Employee training – 77.0%
- Management credibility – 77.2%
- Goal setting – 78.0%
- Hazard correction – 78.0%
- New employee orientation – 79.8%

Overall, companies are relatively strong in the following categories:

- Communication – 80.5%
- Safety contacts – 82.0%
- Accident investigation – 85.8%

Comparison of these data to the criteria for safety success reveals that the categories of recognition, discipline, supervisory training, quality of supervision, and inspections, which are activities typically carried out by supervisors or teams (element #1 made to happen by element #2), would average a score of 66.7% positive, or barely above the red-flag level. This suggests that hourly employees believe their supervisors either do not know how to satisfy their safety responsibilities or that there is no system that requires them to do so.

The categories of management credibility, support for safety, goal setting, and operating procedures, which are means of judging upper and mid-management (elements #2 and #3), average a score of 75.6% positive. Element #4 (employee involvement) scores 74.6% positive.

Differences in Perception

The scores cited above reflect results of hourly employee surveys only. In all cases, supervisors and managers are also polled, and the differences in perception are computed. Within a good organization, the difference is typically 10 to 12 percent; a similar difference exists between managers or executives and hourly employees. A composite picture shows considerably wider discrepancies in these categories:

- Employee training: Supervisors think they are 31 percent better than employees think they are.
- Quality of supervision: Supervisors think they are 25 percent better than employees think they are.
- Inspections: 25 percent better.
- Supervisory training: Nearly 25 percent better.
- Accident investigation: 18 percent better.
- Hazard correction: 17 percent better.
- Attitude toward safety: 14 percent better.
- Support for safety: 13 percent better.
- Communication: 13 percent better.
- New employee orientation: 13 percent better.
- Management credibility: 13 percent better. (Petersen, 2004)

These results suggest that many companies have in some way missed the boat on safety. Historically, firms have taken what can be called an *islands-of-safety approach.* To comply with laws and standards, they create various programs (*islands*), such as a process safety program, a lockout program, a fall protection program, a HazMat program, or an ergonomics program. By creating these islands, however, a company establishes no main channel of solid *management* performance, where everyone from CEO to first-line supervisor takes some action each day that reflects safety as a core value.

As a result, many people come to believe that these "islands" are "safety." They are not. These programs are important components of the overall safety effort, but when a company believes that they satisfy corporate safety responsibility, trouble is on the horizon.

As these data reveal, many safety efforts have lost their focus. Their true focus should be integrated safety, not individual programs that staff can create, thus relieving the line organization of its responsibilities.

Safety excellence only occurs when supervisors, managers, and executives demonstrate their values through actions and by asking hourly workers to help improve the system. This requires a daily proactive approach by line managers and supervisors; a missing link that can only be corrected when the system holds these managers, supervisors, and executives accountable.

Research and benchmarking clearly indicate where safety performance should be. Surveys reveal where performance levels actually are. As these data show, a large discrepancy exists between the two.

Perception surveys give a picture of how safety systems are perceived by the people they try to influence—the workers. Why then are they used so seldom? Dave Johnson, editor of *Industrial Safety & Hygiene News*, discussed this on the publication's Web site:

> So what's the hang-up with perception surveys? Safety pros cite many problems—they are cumbersome, time-consuming, a hassle to administer. Employees have overdosed on one too many workplace questionnaires. Not much forethought goes into survey questions and design. Reasons for probing perceptions are not clear. And what do you do with the results, anyway?
>
> "Ask what action items were identified from their last perception survey and you'll probably get a blank stare," says veteran safety manager Bob Brown.
>
> "People never do anything with the results," says psychologist Dr. E. Scott Geller. "They're never analyzed. Never used for interventions. I visited a client who had gobs and gobs of paper, reams of paper from a perception survey. He looked at me and said, 'Now what?' "

But you can really boil resistance to perception surveys down to one word: FEAR. Results can be too revealing. The emperor is not wearing his PPE. Managers and supervisors are not walking the safety talk. Rules are not enforced. Hazards are not fixed.

"Many times answers can be very threatening to people who think they are doing a good job," says safety consultant Chip Dawson. "People get defensive."

Or it's the fear that the fix is in. That perception surveys are popularity contests. Employees reward friends and punish enemies. Results are rigged to exact a measure of revenge when labor-management relations are lousy.

"Management can perceive that labor uses these surveys to game the system," says Gary Rosenbloom, a risk manager.

But a good argument can be made that now is precisely the wrong time to bury the corporate head in the sand. "If employees are afraid, or feel unsafe or unprepared, shouldn't management know that?" asks Gary Rosenbloom. If managers ignore what they don't want to see, they won't know that the glue that holds workplaces together—things like loyalty, teamwork, and communication—might be melting away.

That lesson comes from the studies of Rensis Likert. Likert pioneered the use of perception surveys to establish relationships between employee attitudes, motivation levels, and how a business performs bottom line. His research connected positive feelings about supportive relationships, group decision-making, and quality of supervision to sales grown, high profitability, and high return on investment.

Here's a thought: Conduct a safety perception survey as the first step to reaching broader business goals. Says safety management consultant Brooks Carder: "After three years of work at a large chemical company, the president remarked that the survey-based safety improvement process had created more positive change in the company's culture than multi-million-dollar engagement by a major consulting firm that took place at the same time."

Here are steps you can take:

Form a perception survey work group consisting of management and employee reps. Be up-front about those fears mentioned earlier, point out the hazards of sticking your head in the sand, and highlighting the work of Deming, Likert, and the Gallup Organization. Do a Google search using "safety perception survey" to find case studies.

Survey a sufficient number of people to ensure statistical validity. Make sure it's a cross-section of all levels of your organization. Jim Stewart, a safety management consultant suggests that for a workplace with 300 people, the survey might include about 70 people (6–12 senior managers, 10 or 20 supervisors, and 50 workers).

Make sure employee participation is anonymous. Use credible, respected co-workers—your "social influencers" and salespeople—to help market the survey and give some positive word of mouth.

Publicize your survey results. Feed results back especially to those employees who took the survey.

Assemble work groups for problem solving. These can be focus groups of employees, supers, and managers that crunch numbers and interpret scores. Develop an action plan with timelines. Plans must be reviewed by senior managers, and implemented with their clear support. Managers must hold assigned individuals accountable for implementing the plan and meeting deadlines.

Monitor what happens next. By all means follow up. Chart changes using activity measures (number of training sessions conducted, hazards corrected, etc.) and repeat the perception survey every year or two.

Most likely, perception survey results will direct your attention to strengths and weaknesses in essential areas of your safety and health program:

1. Management's demonstrated commitment to safety
2. Education and knowledge of workers

3. Effectiveness of the supervisory process
4. Employee involvement and commitment
5. Positive recognition and reinforcement of safety activity. (D. Johnson, 2003)

Measuring Improvement of Safety-System Elements

Different people and groups have identified the essential components of a safety system: the National Safety Council has identified nine elements; NIOSH, five elements; and the British Safety Council, thirty elements. For our purposes, we will identify nine essential elements. They are:

1. Management's credibility, including:
   a. Support for safety
   b. Employee perception of culture
   c. Goal setting
   d. Operational procedures
   e. Employee perception of management credibility

2. Supervisory performance, including:
   a. Quality of supervision
   b. Supervisory training
   c. Employee recognition
   d. Discipline (coaching)
   e. Safety contacts
   f. Inspections

3. Employee involvement and participation

4. Employee training, including:
   a. New employee orientation

5. Employee attitudes and behavior

6. Communications, including:
   a. Awareness programs

7. Accident investigation procedures

8. Hazard control

9. Stress

A safety system's effectiveness is often measured solely on the basis of its failures. Failure-focused measures, such as the lost-time incident rate or the lost-time severity rate, only indicate that problems exist; they do not identify what the specific problems are or how to solve them.

The ultimate goal of any safety system is to prevent or minimize failures. The programs that achieve the goal—the really effective safety systems—focus on activities that are positive, proactive, and designed to target the underlying causes of failures. Activities such as job safety analyses, job safety observations, safety inspections, safety meetings, and safety training are proactive approaches that have demonstrated they promote improvements in overall performance for a safety management system. But many of these are traditional, often perceived as old-fashioned, approaches to safety. In recent years, new approaches to system improvement have been used by management (e.g., Total Quality Management, Statistical Process Control, Six Sigma, etc.). Many of these are applicable to safety management, and are being used there successfully (see Appendix C).

METRICS FOR THE NINE SAFETY COMPONENTS

To ensure continuous improvements in each of the nine areas, measures that will identify progress, or lack of progress, on a regular basis are needed. Some of these will be activity measures, some will be results measures; some will be leading indicators, since they will "lead" us into making corrections; others will serve to fine-tune the element.

Here are some suggested metrics for each of the nine:

1. **Management's credibility.** This category includes:
    a. Support for safety
    b. Perception of culture
    c. Goal setting
    d. Operational procedures
    e. Employee perception of credibility

The key metrics here are the *audit* and *perception surveys*. Other metrics include:

- Percentage of S&H meetings attended by leadership
- Percentage of incident reports reviewed by leadership
- Number of S&H tours taken by leadership
- Percentage of managers with specific S&H roles
- Number of safety goals identified in performance appraisals
- Number of safety goals met in performance appraisals
- Percentage of safety goals that are met on schedule
- Percentage of sampled employees who can state goals and status
- Percentage of work groups with S&H implementation plans

2. **Supervisory performance.** This category includes:
    a. Quality of supervision
    b. Supervisory training
    c. Employee recognition
    d. Discipline (coaching)
    e. Safety contacts
    f. Inspections

    The key metrics in this category are *performance to goal* of the activities for each person in management (a roll-up of percentage to goal at the unit level might give excellent regular, upstream, leading data) and *perception surveys*, which are critical here.

This element clearly requires regular (daily, weekly, monthly) metrics to drive supervisory-to-managerial proactive performance. It requires a system where supervisory and managerial roles are clearly defined, activities to be performed are spelled out, completion of activities are crisply measured, and the percentage to goal of completion flows routinely into both the daily numbers game and the performance appraisal system.

3. **Employee involvement and participation.** The key metric is the *perception survey*. Other metrics include:

   - Percentage of safety surveys completed on time
   - Number of hazard surveys by employees
   - Number of JSAs completed by employees
   - Number of items on action plan owned by employees
   - Number of peer observations made
   - Number of suggestions received per employee
   - Number of employees with stated performance goals
   - Number of improvement teams
   - Percentage of participation in safety meetings
   - Percentage of incidents reviewed by employees
   - Percentage of employees with specific S&H goals and measures
   - Number of recognitions given to individuals

4. **Employee training.** This category covers new employee orientation and current employee training. The key metric is the *perception survey*. Other metrics include:

   - Percentage of training issues identified through hazard analysis, audits
   - Incident investigation
   - Percentage of training courses completed on time
   - Number of employees trained divided by the total number required to be trained
   - Hours of S&H training per employee per year

- Percentage of employees trained within targeted time frame
- S&H training budget per employee
- Percentage of employees passing tests of training effectiveness
- Number of training modules as determined through needs analysis
- Percentage of training modules reviewed or updated per year
- Number of employees trained for cause divided by the total trained
- Percentage of workplace meetings where safety is discussed
- Percentage of design staff with S&H training
- Percentage of design staff with ergonomics training
- Percentage of design staff utilizing ergonomic and safety design tools

5. **Employee attitudes and behavior.** The key metric is the *perception survey*. Other metrics include:

   - Percentage of safe behavior from sampling
   - Percentage of participation in making observations
   - Percentage of observations completed on schedule
   - Number of positive and negative reinforcements given over time
   - Number of groups doing observations divided by the total number of groups
   - Number of trained observers divided by the total number of observers
   - Number of meetings where results are communicated
   - Number of unsafe behaviors observed divided by the number of feedbacks given

6. **Communications.** This category includes awareness programs. The key metric is the *perception survey*. Other metrics include:

   - Number of safety meetings
   - Number of focus-group meetings

- Number of communications to inform
- Perception of effectiveness of incentives

7. **Accident investigation procedures.** The key metrics are the *perception survey* and the *audit*. Other metrics include:
   - Percentage where system causes are identified
   - Percentage where causes of human error are identified
   - Percentage of incident reports that are shared with other units
   - Percentage of follow-up actions and learning shared
   - Percentage of incidents investigated to root causes
   - Average time from incident to investigation completed
   - Average time from incident completion to correction
   - Percentage of investigations that show planning failure
   - Percentage of accident reviews with leadership participation

8. **Hazard control.** The key metrics are the *perception survey* and the *audit*. Other metrics include:
   - Percentage of projects or changes that undergo safety review at:
     - conceptual stage
     - design stage
     - pre-start-up stage
     - post-start-up stage
     - post-start-up verification
   - Percentage of incident investigations shared with planning
   - Percentage of design staff with S&H training
   - Percentage of design staff with ergonomics training
   - Percentage of materials received with MSDS
   - Percentage of materials received verified as meeting specifications
   - Percentage of incidents that show planning failure
   - Percentage of reduction in retrofit costs
   - Completed corrective actions divided by scheduled corrective actions

- Number of corrections implemented divided by the number of deficiencies identified
- Percentage of action items completed on time
- Number of suggestions responded to divided by the number of suggestions received
- Percentage of follow-up or corrective actions and key learnings shared
- Number of best-practice reports prepared per site
- Percentage of serious accident reports that are written up and distributed
- Percentage of facility or business reports related to safety
- Number of suggestions received per employee
- Number of suggestions implemented divided by the number of suggestions received
- Average length of time from receipt of suggestion to response
- Percentage of hazards fixed on schedule
- Percentage of tasks with JSAs
- Percentage of workers involved in hazard identification
- Percentage of first-aid treatments and near-misses investigated
- Average age of outstanding hazards
- Frequency of hazard inspections
- Frequency of updating inspection check sheets
- Number of workers trained in hazard inspections
- Percentage of items fixed versus repair schedule
- Number of emergency drills held
- Evacuation speed
- Percentage of employees trained in emergency procedures
- Average time from receiving a suggestion to responding
- Average age of outstanding recommendations
- Percentage of projects that have safety planning in the budget
- Percentage of purchasing contracts with S&H specifications
- Number of incident investigations that are shared with the planning organization

- Percentage of materials received that have a positive materials identification sheet
- Percentage of materials verified as meeting specifications
- Number of S&H improvement suggestions made divided by the number reviewed and responded to

9. **Stress.** The key metric is the *perception survey*. Others include:
   - Number of downsized employees
   - Average overtime
   - Percentage of employees involved
   - Number of employees in wellness programs
   - Number of stress audits
   - Percentage of employees in stress training programs
   - Number of support groups
   - Percentage in employee assistance programs (EAPs)
   - Level of absenteeism
   - Percentage of alcohol and drug users

There are obviously many measures from which to choose in each category. Each individual organization must choose those that are best for that organization. In making a choice, the criteria established earlier should be used.

References

Bailey, C. *Using Behavioral Techniques to Improve Safety Program Effectiveness.* Washington, D.C.: Association of American Railroads, 1988.

Diekemper, R. and D. Spartz. "A Quantitative and Qualitative Measure of Industrial Safety Activities," in *Self-Evaluation of Occupational Safety & Health Programs*. Washington, DC: NIOSH, 1979.

Farabaugh, P. "OHS Management Systems: A Survey," *Occupational Health & Safety*, March 2000.

Fletcher, J. *The Industrial Environment.* Willowdale, Ontario: National Profile Limited, 1972.

Johnson, D. "Perception is Reality," ISHN E-News (www.ishn.com), vol. 2, no. 30, October 3, 2003.

Johnson, W. *MORT—The Management Oversight Risk Tree.* Washington, D.C.: U.S. Government Printing Office, 1973.

Likert, R. *The Human Organization.* New York, NY: McGraw-Hill, 1976.

Petersen, D. "Safety Management—Our Strengths and Weaknesses," *Professional Safety*, December, 2004.

Planek, T. and K. Fearn, "Re-evaluating Occupational Safety Priorities," *Professional Safety*, October 1993.

Rinefort, F. C. "A study of some of the costs and benefits related to occupational safety and health in selected Texas industries." Ph.D. diss., Texas A&M University, 1976.

Tye, J. *Management Introduction to Total Loss Control.* London: British Safety Council, 1970.

CHAPTER 9

Measuring Company-wide Results: Injury Data

MEASURES OF COMPANY-WIDE SAFETY PERFORMANCE can give us information concerning the company's internal performance, or allow us to compare the performance of one company with that of other organizations. We shall look at both these aspects of measurement in this chapter. First, we will consider traditional measures of company-wide safety performance.

OSHA INCIDENT RATES

Traditionally, we have used one figure to measure company-wide safety performance: the OSHA incident rate, which is the number of injuries per 200,000 worker-hours. For years this has been governed by the rules under OSHA. The United States Bureau of Labor Statistics (BLS) has also judged our national progress with the same measures for some time. Each year it publishes the national rates, giving us a picture of the progress we have made in safety in the country as a whole.

While this is a common measurement, the OSHA incident rate should never be the only measurement of safety performance. Incidence rates take too long to move up or down to provide a meaningful measure from a perspective of immediate feedback. If safety performance at the line level is what prevents incidents, then measuring that performance is the key. However, injury rates do have a place, and as such need to be used as one of a range of measures.

As indicators of internal performance, these rates have some serious weaknesses. The validity and reliability of these measures are a function of the size of the company, or how large the database is, but the real weakness is that the rates are not meaningful to people in the organization. Those in top management often do not understand the rates and may wonder why their safety people cannot talk like managers and use more meaningful measures. As indicators that can be used in comparing the company's progress with that of other organizations, the major difficulty lies with input integrity. Each organization seems to compile its safety records according to its own rules, regardless of the details of the standard. It is common knowledge that, when a company attempts to compare its progress with that of other organizations using these measures, it has no idea how the other organizations have kept their records. Many organizations do not keep their records according to the standard and have no intention of doing so. Still, frequency and severity rates are the final measures used in most record systems today.

The advantages of maintaining these rates include:

- The rates can be compared by standard industrial classification (SIC) codes, and there are comprehensive databases maintained by both the National Safety Council and the Bureau of Labor Statistics.
- The rates provide an even playing field on which to compare departments, plants, or divisions.

The disadvantages include:

- These rates are a result and not a cause of losses.
- They have little meaning to managers, supervisors, or employees.
- Minimization efforts precede rate reductions by one to two years and provide no "real-time" feedback.

- Statistical fluctuations may lead to incorrect conclusions.
- There must be a large number of incidents to attain statistical meaning.
- There is little input integrity from the "other" companies.

OTHER AVAILABLE MEASURES

Since incident rates are not very valuable as measurements, we ought to examine other available measures, such as the following:

1. **Total cost of first-aid cases.** In this appraisal the pro-rata cost per case handled in-plant would be considered. For example, suppose that a first-aid unit in a plant costs $10,000 to operate. Forty percent of the unit's time is used to treat first-aid cases. An average of 1,000 cases are treated; therefore, the cost is $4 per case. This example combines industrial and non-industrial first-aid cases. If average time per case is substantially different for each of these categories, a separate average cost per case may be better.

2. **Total cost incurred.** This includes the actual compensation and medical costs paid for accident cases that occurred in a specified period, plus an estimate of what is still to be paid for those cases.

3. **Estimated cost incurred.** This is an estimate of total cost incurred, based on averages.

4. **The cost factor.** This equals the total compensation and medical cost incurred (see item 3 above) per 1,000 worker-hours of exposure:

$$\text{Cost factor} = \frac{\text{Cost incurred} \times 1{,}000}{\text{Total worker-hours}}$$

5. **Insurance loss ratio.** This is equal to the incurred injury cost divided by the insurance premium:

$$\text{Loss ratio} = \frac{\text{Incurred costs}}{\text{Insurance premium}}$$

6. **Cost of property damage and public liability costs.** This is a measure of damage to property of others caused by company operations.

7. **Nonindustrial disabling injury rate.** This is a measure of off-the-job safety:

$$\text{Injury rate} = \frac{\text{No. of injuries} \times 1{,}000{,}000}{312 \times \text{no. of employees}}$$

The 312 is computed as follows:

$$\begin{aligned} 7 \times 24 \text{ hours} = {} & 168 \text{ hours per week} \\ & \underline{-40} \text{ hours of work} \\ & 128 \\ & \underline{-56} \text{ hours of sleep} \\ & 72 \text{ hours exposed} \end{aligned}$$

$$4.33 \times 72 \text{ hours per week} = 312 \text{ hours per month}$$

The above list is only a beginning. Each organization should devise an injury record system that will measure what management wants measured.

All of the above injury statistics come from results measures; therefore they suffer from the same weaknesses statistically as dollar measures.

Measuring in Dollars

Many of the measures discussed above are dollar-oriented. For internal use, dollar-related measures are perhaps the best. Dollars are understandable and meaningful to everyone in the organization, particularly those at the corporate level. When we talk dollars, we are talking management's language. What dollar indicators work best when dealing with management? The following are some possibilities:

1. Dollar losses (claim costs) from the insurance company
2. Total dollar losses (insurance direct costs) plus first-aid costs not paid by insurance
3. So-called "hidden costs"
4. Estimated costs
5. Insurance loss ratios
6. Insurance premiums
7. Insurance experience modifications
8. Insurance retrospective premiums

DOLLAR LOSSES

The dollar losses outlined in items 1 and 2 are good indicators. (It seems more realistic to include first-aid costs than to ignore them.) Many companies use these figures, or some measurements based on them, such as cost per worker-hour.

Actually, of course, the direct cost of accidents (claim cost) is not money paid out directly by the company. It is money that the insurance carrier pays to the injured employee. Hence, these figures are in some sense *unreal* to management. In addition, they are not easily or accurately obtained. It may be years before actual costs of serious losses are known. Furthermore, using such figures for multiplant companies involves unfair comparison, since compensation benefits vary widely in different states.

HIDDEN COSTS

Hidden costs are real. They consist of such items as:

- Time lost from work by an injured employee
- Time lost by fellow workers

- Loss of efficiency due to breakup of crew
- Time lost by supervisors
- Cost of breaking in a new employee
- Possible damage to tools and equipment
- Time lost while damaged equipment is out of service
- Rejected work and spoilage
- Losses through failure to fill orders on time
- Overhead cost while work is disrupted
- Loss in earning power
- Economic loss to employee's family
- At least 100 other items of cost that may arise from any accident

Although these costs are very real, they are difficult to demonstrate. To say, arbitrarily, to management that they amount to four times the insurable costs is asking for trouble. If management asks for proof, you can only say, "Heinrich said so." Management wants facts—not fantasy. Without proof, hidden costs become fantasy.

ESTIMATED COSTS

The actual direct cost of a work injury, which includes compensation and medical benefits, often is not established until long after the injury occurs, especially for more severe cases. Thus, the loss statements provided by insurance carriers do not serve the purpose of effective, current cost evaluation. The only information that such statements can provide on relatively recent injuries consists of the amounts of medical and compensation benefits paid to date, plus the outstanding reserves. These reserves are established primarily to assure that sufficient funds are set aside for the eventual cost of the claims. It is not until the healing period has ended and the degree of disability has been positively determined, however, that any accurate estimation of the cost can be made. And the more complicated the injury, the longer it takes to establish the final cost.

In recent years, there has been an increasing demand for some method by which employers can determine promptly the approximate cost of compensable occupational losses.

Management pays an insurance premium for worker's compensation coverage. There is a real cost to it. How much it pays is directly dependent on the company's past and present accident record. This premium is paid directly out of management's pocket. Why not then measure the results in terms of this real out-of-pocket cost to management? The insurance premium is based on industry averages, adjusted by each company's record in the past and, in some cases, adjusted again by the current accident record. These adjustment factors, known as the *experience modifiers* and the *retrospective adjustment*, are excellent indicators of past and present performance. Safety people ought to know, understand, and use these adjustment modifiers, since they represent the true costs of safety to management.

WORKERS' COMPENSATION INSURANCE COSTS

There are both direct and indirect costs associated with worker injury incidents (covered losses): direct costs are those medical and compensation costs paid to the claimant by the insurance company; indirect costs are the so-called "hidden" costs not covered by insurance and, in fact, are not easily observed or recorded. Indirect costs include time lost by others who observed or gave help at the time of the insured incident and time lost by the supervisor in investigating.

Many studies have been made (by Heinrich, Bird, etc.) of indirect costs (damaged products, loss in production time, etc.), with varying results. Each incident that occurs will, of course, have a different indirect cost and a different indirect-cost-to-direct-cost ratio. In some industries, particularly some manufacturing sectors, this ratio may be consistent and useful. Obviously, the ratio will vary tremendously, depending on circumstances, injury severity, and other factors.

More important, any indirect costs that a safety specialist claims on the basis of some assumed ratio, even on the basis of computed past figures, will be somewhat questionable, and management knows this.

To state arbitrarily that direct costs are $10,000, indirect costs are $40,000, and total incident costs are therefore $50,000, usually is not good for much more than safety preaching. The reality may simply be that the system was inefficient, productive time was lost, and quality was adversely impacted.

Direct costs are more easily observed and measured. They are close to being real costs—in fact, they *are* real costs if not clouded by insurance company "reserves." Reserves constitute the insurance company's final estimate of dollar losses per case, and they must be established for serious cases. A safety professional whose estimate shows these reserves to be inaccurate can, of course, establish an estimate with his or her own figures on those cases still open, provided that management understands this approach.

Your insurance company may reserve a broken-arm case at a specific dollar amount based on their experience. If your company has a good return-to-work process you might expect to pay half of your insurance company's projection. On the other hand, if your injured employee's job is very physically challenging, it may take longer for him or her to return to work, so you may want to leave the reserve at the existing level, and only increase it if you are sure it will be low. Remember: reserves have a big impact on a company's insurance costs.

Even direct costs are somewhat *unreal* to the average company since they are paid by the insurance carrier, not by the company itself. However, the amount the company does pay out is dependent to a large extent on what these direct costs are, on the insurance contract, and on whether a company is self-insured. (Some contracts specify a guaranteed cost based on the retrospective maximum.)

This discussion is included to give the reader a better understanding of how the insurance premium is arrived at, thus allowing them to use this premium in communicating with management. We are referring here only to the determination of the workers' compensation premium, since this premium is of the greatest interest to most safety professionals. (However, an organization's safety performance can also affect automobile and general liability insurance.) Also, in most states this premium figure is not as subject to manipulation as premiums on other lines of insurance, making it a more objective measure of safety performance. The workers' compensation premium is determined by three rating systems:

1. Manual (or state) rating
2. Experience rating
3. Retrospective rating.

In devising rates, the first step is to determine a basis of exposure. In workers' compensation, that basis is payroll. This reflects the number of employees, the hours they work or are exposed, and their pay scale. It is a measure of exposure that can be more readily ascertained and more easily verified than any other.

Workers' compensation rates are expressed in terms of dollars per $100 of payroll. Different types of work entail various degrees of hazard; therefore, in rate-making, the types of work are *classified* to give consideration to these degrees of hazard. Through long experience and study of injury records from all types of businesses, a list of classifications has been developed in which the different types of work are arranged according to the degree of hazard. There are 600 to 700 classifications. Each company fits into one or more of these.

Of course, each company must first make sure it is in the correct classification, as each classification has its own rate. Each classification, or group of similar industries, actually sets its own insurance rate.

Each year, all insurance carriers report by classification to the state and to the National Council for Compensation Insurance the payroll of each company they insure and details of losses incurred by each company for the year. Then, on the basis of what happened to all companies in the classification during the last five years, a new rate is computed each year. Manual rates thus are governed by the experience of the industry over the last four years, not including the immediate past year.

Next, an experience rating is considered. It should be pointed out that although the experience-rating process described below is the one used in the majority of states, it may not be used in your state. Not all states follow this process exactly; many elect to use a somewhat different format.

After manual rates are applied, experience rating is used to vary the company's own rates, depending on its experience in recent years. Under the manual rate, all industries of the same type (such as machine shops) pay exactly the same rate, regardless of their own workers' compensation claims record. Obviously, in any group of machine shops, some will have a good safety record, and some will have a poor one.

Experience rating attempts to change this so that the company with a good safety record pays less than the company with high claims' losses. Experience rating makes a statistical comparison between what losses *occurred* in a particular company during the past three years and what losses *were expected to occur* during that period in a shop of that size.

Part 1 of the experience rating form merely lists all incidents that have occurred in a company during the last three years. This is done as follows: First, all claims that cost less than $XXX each are lumped together. Example:

Year 1: 15 incidents, costing a total of $4,563
Year 2: 10 incidents, costing a total of $2,288
Year 3: 10 incidents, costing a total of $1,404

Next all incidents that cost over $XXX each are listed individually:

Year 1: one case still open (O) at $7,287
Year 2: one case now closed (F) at $2,587
Year 2: one case now closed (F) at $876
Year 3: one case still open (O) at $953
Year 3: one case now closed (F) at $789

These more expensive cases are then discounted. They go into the rating at a lesser amount. For example, the open case in Year 1 at $7,287 is discounted to $2,656; the same is true for the other cases.

The actual discount applied is based on a statistical formula, but the larger the loss, the larger the discount. There are several reasons why the large losses are discounted in the rating plan:

1. Without this discounting, one large loss would make the cost of insurance prohibitive for the smaller company.

2. The plan states in effect that high frequency of losses should be penalized, for it indicates poor management of loss control.
3. The plan says that one serious incident is less indicative of a poor insurance risk than a batch of small ones.
4. The rating structure is still built around the theory that severity is fortuitous.

In part II of the experience rating form, expected losses are computed. First, classifications are listed. (Each classification describes a type of operation; in this case, classification 3400 is metal goods manufacturing, classification 8810 is office employees, and classification 8742 is outside salespeople.) Next, the annual payrolls are listed, and then an expected loss rate for each classification is indicated. The expected loss rate is 60 percent of the manual rate for the classification, and this rate times the payroll shows how many losses would be expected for that classification for that period. In the three years in our example, we would expect the metal goods manufacturing portion of this plant to incur $32,847 in losses.

We totaled the actual losses and the expected losses for the company. We also discounted both. In the rating form, we compare the resultant two figures. In this case, our figures are $15,265 actual losses and $18,643 expected losses. This company has a better-than-average record.

Now ask the question, "Is this believable? This is what actually happened, but could we expect it to happen again and again?" To answer this question of believability, several other actuarial factors are added.

RETROSPECTIVE RATING

After the experience rating is applied, the company may select a retrospective workers' compensation rating plan of insurance instead of a regular plan. In a retrospective plan, the amount of premiums paid will depend on the amount of losses that occurred this year (not in the past three years). Retrospective adjustments are made on top of, after, or in addition to the experience rating.

The experience modification cannot be escaped by the company; it must pay that rate. In our example it pays 82 percent of the average, or manual, rate. (This is sometimes called a credit.) With a poor loss record it could have paid 110 or 200 percent of the manual rate; this is sometimes called a debit. When this occurs, some companies try to lower insurance costs by opting for a retrospective rating.

Choosing a retrospective rating is a gamble; a company decides on partial self-insurance or, rather, on cost-plus insurance. The premium paid under this plan depends on the losses the company will sustain this year. In retrospective rating, the company will pay for (1) the administration of the insurance company; (2) its losses; (3) the use of the insurance company's claim department; and (4) taxes, which are all subject to a minimum and a maximum.

The higher the losses during the year, the higher the insurance premium will climb. If management succeeds in keeping losses low, the insurance premium will remain low. At the same time, it must pay a penalty if its losses go over the break-even point. This is shown in Exhibit 9.1. Obviously, under retrospective rating, management becomes quite interested in loss control.

At the beginning of the policy year, management decides whether it wants retrospective rating and, if so, what kind of special plan would be best. There are a number of retrospective plans. The slope of the line, the minimum and maximum premiums, the break-even points, and other factors, all vary with the different plans.

■ Exhibit 9.1 **Retrospective rating**

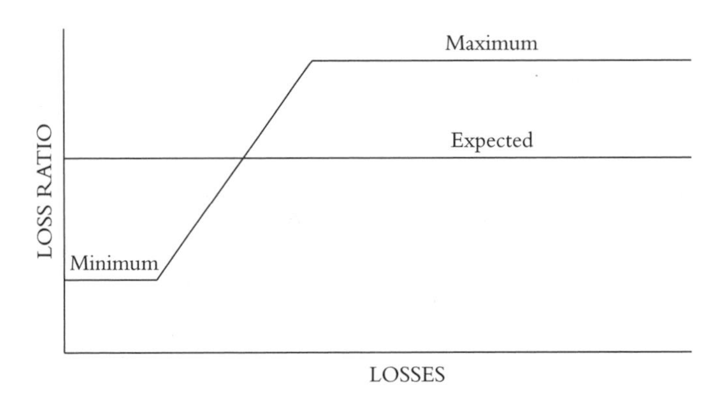

As safety specialists, we should be aware of these insurance costs and of the different plans that set the costs. We should know our manual rate, our experience modifier (EMR) (credit or debit), and whether the company is on a retrospective rating. Management looks at these insurance costs. Safety people ought to be very familiar with these figures and should utilize them in their communications with management.

Thus, many measures can be used to judge corporate progress and performance in safety. There are both trailing and leading indicators or metrics. Results (failure) measures such as dollar-oriented measures, insurance measures, and estimated cost measures can be used, as well as before-the-fact results measures such as inspections and safety sampling. OSHA code violations, discovered either by the safety specialist or by compliance officers, can also be used as measures. In short, if we want to judge a firm's internal safety performance, there are a wide variety of measures from which to choose. It is hoped that safety professionals make informed selections based on some of the criteria outlined earlier in this chapter.

In point of fact, however, top executives rather than safety staff often choose the measures that will be used. There is nothing wrong with this, but safety staff should encourage corporate executives to find more meaningful measures than the traditional (trailing) ones.

Beyond mapping out a company-wide gauge of safety performance, it is even more difficult to find viable measures for comparing one company's progress with that of others. Most safety professionals agree that traditional methods of doing this are ineffective and inaccurate.

However, the real question appears to be whether we need such measures. Do we really need to compare ourselves with others? Is such a comparison meaningful? A company's safety record reflects hazards, controls, employee morale, the climate and style of the company, and many other things. It is difficult to see how or why a company should compare itself with others when all the items that go into shaping each company's record are different. It would seem that the only important thing we need to know is whether we are getting better or worse in each period measured.

Workers' compensation costs seem like a natural measure—and they are—but understanding the data limitations is important.

The advantages of worker-compensation data include:

- They provide a monetary basis for the measurement of efforts and success.
- With organization and appropriate tools, they can provide the line organization with meaningful data.
- Workers' compensation costs allow for budgeting and the charge-back of compensation costs. Departments pay for their losses and not other groups' losses.

The disadvantages include:

- The use of concepts like "reserves" and "paids," as well as the time to clear for claims, which are difficult for some managers to accept.
- The necessity to maintain knowledgeable staff personnel in the compensation area to work with safety, the carrier, and the broker.

FINANCIAL METRICS

The goal of financial metrics is to answer the following questions:

1. What contributions do safety, health, and environmental (SH&E) initiatives make to the "bottom line"?
2. Are SH&E goals/needs in line with company financial goals?
3. Does company management understand the financial impact of SH&E decisions? (Some organizations have demonstrated a significant positive impact on both productivity and quality.)

The need for financial metrics is being driven by external and internal competitive pressure and competition for internal funding. These forces make understanding the safety and health contribution to business performance of growing importance for the continuing success of such programs. It is vital for the safety and health professional to speak the language of business—the financial impact in terms of productivity; quality and, ultimately, dollars are understandable and meaningful to everyone in the organization, particularly those at the corporate level.

Safety and health professionals should gain an understanding of costs and cost-drivers and know what can and cannot be impacted in the organization. Based on this knowledge, they should determine what activities to change, work toward improving them, and then measure progress. Activity effectiveness can be measured in a number of ways, including comparing real costs to budget, to the prior year, as a percent of net sales, or to an industry standard. The objective should be to find the financial measure that drives management to take action. Cost analysis must extend beyond incidents and their direct costs to incorporate the full impact of safety on the business. For example, what is the cost of a production line going down in terms of lost market share?

Measures of Company-wide Results

The *preferred* metric is: value of injury- or illness-related losses in terms of sales required to be recouped. Other metrics include:

- Budget for SH&E staff, training, and other needs over time
- SH&E budget: dollars per employee
- Cost per employee of SH&E prevention—actual dollars spent per employee
- Number of SH&E personnel divided by the total number of employees
- Injuries per employee hours
- Number of zero days
- Average loss per product line; average gain per product unit
- Total losses over time (includes products, property, liability, personal injury, business interruptions, inventory loss, etc.)
- Total gains over time per units of production (UOP), per incident (I), per employee (E)
- Significant losses over time
- Cost of incidents per unit of production
- Cost of incidents per employee

- Workers' compensation costs per employee, per hour, per unit of production
- Experience modification factor
- Cost per employee or per unit for SH&E expenditures—prevention + losses (direct and indirect)
- Fines and penalties measured per UOP, per incident, per employee
  production time lost due to employee incident
  quality improvement resulting from improved employee ergonomics
- Accident costs versus industry average costs or competitor costs
- Ratio of occupational to nonoccupational losses over time
- SH&E direct losses over time
- Direct losses per total shares
- SH&E direct losses on earnings per share over time
- Workers' compensation costs per employee over time

References

Heinrich, H. *Industrial Accident Prevention*, 5th ed. New York: McGraw-Hill, 1980.

Bird, F. and R.G. Loftus. *Loss Control Management.* Loganville, GA: Institute Press, 1976.

CHAPTER **10**

Using Scorecards: A Macro Measure

OUR DISCONTENT WITH SUBSETS of incident data as a metric forced us to use other measures years ago, starting with the development of the audit—checking whether designated employees were performing predetermined activities to achieve safety. We perceived audits to be effective measures until we began to run some correlational studies between audit scores and accident statistics in large companies over time. We often found zero correlations, even some negative correlations, between the two.

THE AUDIT AND CORRELATIONAL STUDIES

As far back as the 1970s we were beginning to research the relationship between audits and accident statistics. In a ten-year study by the Association of American Railroads (AAR), an extensive survey was made of safety activities performed by all major railroads in the United States. The report of the survey stated:

> An extensive survey of safety program activities was completed, which covered more than 85% of the employees in US railroads at that time.
>
> A detailed questionnaire was completed by the Safety Office and staff on each railroad. The questions were designed to determine the presence of and absence of activities popularly thought to influence

> program effectiveness and to trace the organization's responsibility, staffing, program content, effectiveness measures, and perceived quality of performance. The information gained was then fed into a computer to be compared with other data measuring safety program effectiveness. (Bailey, 1988)

This activities survey was basically an audit covering the following twelve components of a safety system:

1. Safety-program content
2. Equipment and facilities resources
3. Monetary resources
4. Reviews, audits and inspections
5. Procedures development, review, and modifications
6. Corrective actions
7. Accident reporting and analysis
8. Safety training
9. Motivational procedures
10. Hazard-control technology
11. Safety authority
12. Program documentation

At the same time, the precursor to today's perception survey was developed by the AAR study group, validated by the test development experts, and administered to the same companies.

The results of the audit and of the perception surveys were then correlated with the accident statistic for these large organizations.

The hypothesis was that high scores in the twelve areas would correlate generally with the lower accident and injury rates; instead, they found little correlation with these factors. Their report of findings stated, in part:

> It was an unexpected result of this study that so little correlation was found to exist between actual safety performance and safety activity scores. The overall score has almost no correlation with train accident rates and cost indicators and is somewhat counter-indicative with respect to personal injury rates. The only two categories which correlated consistently and properly with accident rates were monetary resources and hazards control. Two categories: equipment and facilities resources and reviews, audits and inspections, had counter-intuitive correlations. (Bailey and Petersen, 1989)

Thus, results seemed to be saying:

1. The effectiveness of safety programs cannot be measured by the more traditional criteria popularly thought to be factors in successful programs.
2. A better measure of safety-program effectiveness is the response from the entire organization to questions about the quality of the management systems that have an effect on human behavior relating to safety.

Similar studies have been made since the 1970s, correlating audit results with accident statistics. The results usually show little or no correlation. During a recent study of audits at Tulane University, some of the reasons were discussed. The study examined the characteristics of nine different Occupational Health and Safety Management Models or audits. Each was unique. The OSHA model suggested five basic components for a safety system. The American Industrial Hygiene Association suggested five as well, but all five were different from OSHA's model. The British Standards Institute had two basic elements, the Department of Energy had five different ones, the Chemical Manufacturers' model had six; Det Norske Veritas had three, the Hospital Association used two, and the Australian Work covered two. (Farabaugh, 2000)

Some packaged models not mentioned in the Tulane study were the British Safety Council, which at one time used thirty elements; the Nosa International system from South Africa with five; and the original ILC system, which had 17 to 20.

In short, they are all different because they reflect the ideas and biases of the persons who originally put their beliefs down on paper. Some seem patterned after the original concept of Jack Fletcher from Canada, or Frank Bird in his work originally done for the South Africa Mining Industry. Others have copied the original concepts of Roman Diekemper and Donald Spartz.

Seldom do you find any mention of correlational studies between the content of these audits and the accident record.

The whole concept of the audit is that there are certain defined things that must be included in a safety system to get a high rating or the greatest number of stars. The research, however, does not agree:

- A NIOSH study in 1978 identified seven crucial areas needed for safety performance. Most are not included in the above programs. (Diekemper and Spartz, 1979)
- A Michigan state study had similar results.
- Foster Rinefort's doctoral dissertation at Texas A&M suggested there was no one right set of elements. (Rinefort, 1976)
- A 1967 National Safety Council study suggested many of the "required" elements of packaged programs were quite suspect in terms of effectiveness. (Planek, 1993)
- A 1992 National Safety Council study replicated the 1967 study with the same conclusions. (Planek, 1993)
- The study done by the Association of American Railroads conclusively showed that the elements in most packaged programs had little correlation with bottom-line results. (Bailey, 1988)

In light of today's management thinking and research, the audit concept has become suspect. In most systems, a number of elements were defined; and in most, all elements were equally weighted. Thus, having the right books in the corporate safety library counted as much as whether or not supervisors were held accountable for doing anything about safety. For the most part, safety professionals tacitly accepted the components of these packaged audits and never questioned the lack of any graded scale of element value.

In addition, there was apparently little effort to correlate the audit results to the accident record. Thus, safety people were buying into an unproven concept. When some correlational studies were run, the results were surprising:

- One Canadian oil company location consciously chose to lower their audit score and found their frequency rate significantly improved as a result.
- One chain of U.S. department stores found no correlation between the audit scores and workers' compensation losses. They found a negative correlation between the audit scores and public liability loss payout. (Petersen, 1996)

Some organizations have used the audit as a primary metric very successfully because, over time, they have run correlational studies to ensure that the elements in their safety system do in fact get results. One example of this is the Proctor & Gamble system.

Eugene Earnest, formerly Safety Director for P&G, explained their system:

> A number of years ago, the P&G corporate safety group developed what is known as the Key Elements of Industrial Hygiene & Safety. In effect, these are the what counts activities for IH&S and are the basis for site surveys. It was believed that if these activities were effectively implemented, injuries and illness would be reduced and conversely, if they were done poorly, injuries and illness would increase. Because line management was involved in the development of this list, there was "buy in."
>
> I cannot stress enough the importance of having a clearly identified IH&S program against which goals can be established at all levels of the organization and people held accountable for before-the-fact measures of injury and illness prevention.
>
> Each key element is rated by the surveyor utilizing a scale of zero to ten where 0 means "nothing has been done" and 10 means the key element is "fully implemented and effective." (Earnest, 1994)

The validity of the key elements has been proven through correlational studies over the years. The audit can be an effective metric once an organization is sure that the audit elements will lead to real results.

THE PERCEPTION SURVEY AS A METRIC

The perception survey is used to assess the current status of an organization's safety culture. Critical safety issues are rapidly identified and any differences in management and employee views on the effectiveness of company safety programs are clearly demonstrated.

The goal of the perception survey is to understand how a company is performing in each of a number of safety categories. The survey begins with a short set of demographic questions that will be used to organize the graphs and tables that show the results.

The second part of the survey consists of questions that are designed to uncover employee perceptions about the safety categories. Questions have been statistically, individually validated over years of use in a particular field; development of one perception survey was a product of the previously mentioned study for the Association of American Railroads.

As a result of that nine-year study, the following conclusions were reached:

1. The effectiveness of safety efforts cannot be measured by traditional audit criteria.
2. The effectiveness of safety efforts can be measured with surveys of employee (hourly to executive) perceptions.
3. A perception survey can effectively identify the strengths and weaknesses of elements of a safety system.
4. A perception survey can effectively identify major discrepancies in the perception of program elements between hourly rate employees and levels of management.
5. A perception survey can effectively identify improvements in and deterioration of safety-system elements if administered periodically.

The above conclusions, based on the data, carry with them considerable importance to safety management thinking. In effect, the data strongly suggest that the twelve audit elements described as a systems approach are *not* closely related to safety performance or results.

The twelve components are not a bad description of the elements of many safety programs in use in industry today. This data does not question the audit approach to safety performance improvement. It does question the validity of any audit approach made up of arbitrarily decided elements, such as these twelve.

It supports rather strongly the belief that a perception survey as described in this chapter (sometimes called a *culture survey*), properly constructed, is a better measure of safety performance and a much better predictor of safety results.

THE SCORECARD APPROACH

The trend today is toward multiple measures to assess safety-system effectiveness. These usually include at least three measures:

1. The accident record
2. The audit score
3. Perception-survey results.

Currently, the accident record (statistics) is still used because management is extremely hesitant to give it up. Eventually, it may be phased out of the scorecard. Other scorecard measures include:

4. Behavior-sampling results
5. Percentage to goal on system improvements
6. Dollars (claim costs, total costs of safety, etc.).

Proctor and Gamble's scorecard contains two measures: the OSHA incident rate and key element ratings (basically an audit score). MeadWestvaco Corporation is considering using three: incident rate, perception-survey results, and audit scores. PPG Industries is considering using four measures.

Other organizations are experimenting with different mixes for their scorecard of metrics to assess safety-system effectiveness:

- Navistar uses eight: incident frequency rate, lost-time case rate, disability costs, percent improvement in safety performance, actual healthcare costs, absenteeism, short-term disability, and long-term disability. (Navistar, 1999)
- Kodak sets goals and measures in seven areas: lost time, plant operations matrix (percentage to goal), employee surveys, assessment findings, integration matrix, vendor selection, and "best in class" (a benchmark metric). (Esler, 1999)
- The National Safety Council has suggested "performance indexing," which includes six: number of team audits, process safety observations, employee attitude ratings, required safety training, safe-acts index, and management audits.

There is considerable new and innovative thinking taking place in many organizations regarding safety-system content and safety metrics. There is no *one* "right way." Each organization must determine its own right way.

In addition, after deciding the components to be included in the scorecard, a company must also decide how each component should be weighted, making it possible to come up with a single metric, if that is what is desired.

This poses two serious problems: (1) figuring out what should go into the scorecard, and (2) convincing middle and upper management to participate in ranking the scorecard elements. These changes pose serious challenges to the safety professional. It might be necessary to create considerable dissatisfaction with the status-quo metrics in an organization (cognitive dissonance), from the CEO down, and then install the selected scorecard to replace the current single metric. Since this could mean eliminating the executives' way of comparing the company to other organizations, expect considerable reticence to any change at that level.

Nevertheless, at some point it will have to happen. Probably the sooner one starts down this route the better.

References

Bailey, C. *Using Behavioral Techniques to Improve Safety Program Effectiveness.* Washington: Association of American Railroads, 1988.

Bailey, C. and D. Petersen. "Using Perception Surveys to Assess Safety System Effectiveness," *Professional Safety*, February 1989.

Bird, F. *International Safety Rating System*, 5th ed. Loganville, GA: International Loss Control Institute, 1988.

Diekemper, R. and D. Spartz. "A Quantitative and Qualitative Measure of Industrial Safety Activities," in *Self-Evaluation of Occupational Safety & Health Programs*. Washington, DC: NIOSH, 1979.

Earnest, E. "What Counts in Safety," in *Insights into Management*. National Safety Management Society, 1994.

Esler, J. "Kodak's Health, Safety and Environmental Performance Improvement Program," presentation to ORC, 1999.

Farabaugh, P. "OHS Management Systems; A Survey in Occupational Health and Safety," March, 2000.

Fletcher, J. *Total Loss Control.* Toronto: National Profile Ltd., 1972.

Herbert, D. "Measuring Safety and Trends in Safety Performance," presentation at National Safety Congress, 1999.

Navistar. "Operationalizing Health and Productivity Management," presentation to ORC, March, 1999.

Petersen, D. *The Perception Survey Manual.* Portland, OR: Core Media Training Systems, 1993.

Petersen, D. *Techniques of Safety Management*, 3rd ed. Des Plaines: ASSE, 1996.

Planek, T. and K. Fearn, "Re-evaluating Occupational Safety Priorities," *Professional Safety*, October 1993.

Rinefort, F. C. "A study of some of the costs and benefits related to occupational safety and health in selected Texas industries," Ph.D. diss., Texas A&M University, 1976.

CHAPTER 11

Measuring National Results

THE FOLLOWING CRITERIA MIGHT BE USED as a measure of performance at the national level:

1. It should be valid.
2. It should be statistically reliable.
3. It should be a results measurement.
4. It should ensure input integrity.
5. It should be understandable to all.
6. It should be able to be computerized.

Currently, the measures listed below are those in use:

1. Frequency rates
2. Severity rates
3. Fatalities and fatality rates
4. Estimates of costs to the country

How do these four measures meet the established criteria? First, they are (or could be) statistically valid and reliable, for certainly the database is large enough. They are all results measurements. For the most part, they are understandable to those who use them, and certainly they can be computerized. The one criterion that the measures do not meet is that of ensuring input integrity. And that presents a major problem. In actuality, companies are using batches of figures that are often sheer guesses. There appears to be little control over the input of national figures, whether from the National Safety Council or from government. We have detailed a standard for two measures (frequency and severity rates), but that has certainly not ensured input integrity. So it is mere supposition to assume that OSHA's figures are any better at ensuring input integrity.

In this chapter, we look at some of these problems. First, however, we will examine our current measures to see whether they could be altered to ensure input integrity, which is a function less of the measurement itself than of the data-collection system.

CURRENT MEASURES

As mentioned earlier, the current, nationally accepted measures are the OSHA incident rate, the severity rate, the number of fatalities, and estimated cost measures. Measures of fatalities at the national level are understandable; they are results indicators and can be computerized. They do suffer from the lack of input integrity, of course, but even in this respect they seem better than other measures. Asking whether a death count or death rate really reflects our national effort and performance in the safety area raises serious doubts. We seem to believe that fatalities are caused largely by chance, and thus we tend to raise our eyebrows at such a measure. Yet the criteria tell us that measures of fatalities are not too bad; perhaps they are the best of our traditional measures for judging national safety performance, since they seem to be capable of more input integrity than the others.

Cost measures are a different story. While they meet some criteria except, again, that of ensuring input integrity, it seems that cost figures are so far removed from reality that any attempt at using national costs seems to be a futile exercise. For example, the 2001 edition of *Injury*

Facts, published by the National Safety Council, gives these values for the costs of all accidents in the year 2000:

Wage loss	67.6 billion
Medical expenses	24.2 billion
Insurance administration costs	22.3 billion
Property damage (motor vehicle)	2.2 billion
Fire loss	3.4 billion
Indirect cost from work accidents	11.5 billion
Broken down among:	
Motor-vehicle accidents	201.5 billion
Work accidents	131.2 billion
Home accidents	111.9 billion
Public non-motor-vehicle accidents	512.4 billion

What is the source for the above figures? The footnote to this table indicates that the source is NSC estimates (rounded), based on various inputs: national figures; state figures; and figures from insurance companies, industries, and others. How are these figures actually compiled? By piling one batch of estimates upon another batch of estimates. In short, practically no solid, factual input is used to arrive at these figures.

How about incident rates? These figures look so much more solid in the BLS statistics that we think they surely must be better, but they really are not. Besides the input integrity problem, they are somewhat less meaningful to most people than dollars and deaths. But the main problem remains the simple fact that since some people do not follow the standard in putting together the rates at their places of business (the input site), the result, nationally, is questionable.

WHAT SHOULD NATIONAL MEASURES TELL US?

If we expect to apply meaningful yardsticks to safety endeavors on a national level, we must answer the question: What are we trying to measure? Are we trying to measure our collective national safety performance? If so, we should be structuring our measurements along the same lines followed by companies (i.e., we should be asking what percentage of companies in this country have safety policies, what percentage train their people, what percentage hold supervisors accountable, etc.). Perhaps this could be accomplished with profiling systems. We could sample and profile individual companies and in some way combine the results regularly into a profile of U.S. corporations. Our national progress would be measured by the improvement in our collective U.S. corporate profile.

Or are we trying to measure our collective national safety results? If so, we must find a solution to the input integrity problem. Perhaps the answer is to use periodic sampling of selected plants and industries, rather than attempting widespread data collection in which input quality always varies. While sampling covers only a small percentage of companies, statistical approaches can make this percentage quite valid, reliable, and totally reflective of the national picture. More important, even with some degree of error in sampling, I have no doubt that it gives us a more accurate picture than massive data collection in which the degree of error is enormous.

Or are we trying to measure our collective national results before accidents occur? Sampling of behavior or sampling of conditions could provide these data better, with randomly selected industries sampled routinely.

What we currently do to find out where we stand nationally is almost worthless in terms of accuracy and usability.

Measuring OSHA and Other Enforcement Agencies

A related problem is that of measuring the effectiveness of OSHA and similar agencies charged with law enforcement and with improving safety in our nation's industries. OSHA, for example, has been under constant fire since its inception; frequency- and severity-rate indicators have shown that the agency has had little impact on our national safety problem. Are these indicators the best way to measure OSHA? Is this a *fair* measure of its performance? Examined up close, it seems to be a profoundly *unfair* measure.

By law, OSHA is charged with ensuring that employers provide a work environment that is free from recognized hazards that cause or are likely to cause death or physical harm, and that employers comply with the standards promulgated. Thus, if we are to measure OSHA's performance, we should devise a means for measuring:

1. Whether workplaces really have fewer recognized hazards than before the enactment of OSHA.
2. Whether there is actual compliance with the standards.

Since it is generally agreed that the presence of recognized hazards and a lack of compliance with standards account for a relatively small percentage of accidents (around 10 percent, according to Heinrich), it is grossly unfair to charge OSHA with the other 90 percent. Thus, it is grossly unfair to measure OSHA by the overall accident record of this country. If, at best, OSHA was devised to remedy 10 percent of our accident problem, it should be measured by only the 10 percent of accidents with which it is concerned. Better yet, it should be measured by whether workplaces are freer of recognized hazards than they were before its inception and by whether OSHA standards are met.

What specifically should be measured? OSHA itself seems to like its own internal measures: number of inspections made, number of citations issued, and number of fines levied. While initially this seems to make sense, it really is nothing more than a massive numbers game. Results such as these are almost totally under the direct control of the administration. For instance, if you are being measured by number of inspections made, you can improve your position merely by calling on smaller companies, which will enable you to visit more each day. Similarly, if you are being measured by number of violations found, you can easily improve your showing by calling on only a few large companies. If the number of fines is what is measured, just being harder on everyone will make your record look better. These measures have almost no relation to safety and little relation to either standards' compliance or fewer hazards in the workplace.

If measuring OSHA's effectiveness is the goal, the yardsticks to use are exactly its stated objectives: Do workplaces have fewer hazards than before, and are standards being complied with? While the measurement problem is not an easy one, the solution seems to lie in a sampling approach to workplaces—sampling a statistically sound percentage of America's businesses annually and noting progress in terms of number of hazards per employee (or worker-hour), number of violations per worker-hour, and similar gauges.

NATIONAL AND INTERNATIONAL MEASURES

Some safety professionals say that our national measures should also tell us where we stand in comparison with other nations. This would be an almost total waste of time. Different countries have different mores, values, attitudes, highway systems, cultures, job approaches, kinds and amounts of industries, job practices, and physical work conditions. A safety record is nothing more than a reflection of all these things. It seems illogical to attempt to compare one country's safety performance with that of another. It seems most logical, nationally, to compare our present record only with our past record: Are we getting any better?

References

Heinrich, H. *Industrial Accident Prevention*, 5th ed. New York: McGraw-Hill, 1980.
National Safety Council. *Injury Facts*. Itasca, IL: NSC, 2001

CONCLUSION

The Process for Achieving Safety Excellence

ONE OF THE BEST BOOKS ON MANAGEMENT is *Leadership Is an Art* by Max De Pree. His book begins with a profound statement: "The first job of the leader is to define reality." Perhaps this is the single most important thing a leader in safety can do.

Historically our corporate safety leaders (CEOs, COOs, etc.) have not attempted to define reality. It is only in recent years that we have had the tools to do this. Usually we jump to solutions of our own or accept those dictated by others.

Since Max De Pree posited his leadership concept, the road toward safety excellence has been threefold:

1. Define reality—where are we today?
2. Define the vision—where do we want to be?
3. Define the process—how do we get there?

DEFINING REALITY

Defining reality is more possible today in safety than ever before. Until recently, we have depended on injury figures to tell us the *reality* of our safety efforts. There has now been enough written about the invalidity of such measures to demonstrate the ridiculousness of this approach. Even our dependence on audits has been severely questioned in recent years. However, we have developed better upstream measures (based on Deming's philosophies, perception surveys, and other tools) to provide the data needed to define reality.

Safety excellence only occurs when supervisors, managers, and executives demonstrate their values through actions, and their credibility by asking hourly workers to improve the system. This requires a proactive approach daily by line managers and supervisors—often a missing link that can only be corrected when the system holds these managers, supervisors, and executives accountable.

Research and benchmarking clearly indicate where safety performance should be. Surveys reveal where performance levels actually are.

Appendix A

The following is an example of an audit from the National Institute of Occupational Safety and Health (NIOSH).

A. ORGANIZATION & ADMINISTRATION

Activity	Poor	Fair	Good	Excellent
1. Statement of policy, responsibilities assigned.	No statement of Loss Control policy. Responsibility and accountability not assigned.	A general understanding of Loss Control, responsibilities and accountability, but not written.	Loss Control Policy and responsibilities written and distributed to supervisors.	In addition to "Good," Loss Control policy is reviewed annually and is posted. Responsibility and accountability is emphasized in supervisory performance evaluations.
2. Safe operating procedures (SOPs).	No written SOPs.	Written SOPs for some, but not all, hazardous operations.	Written SOPs for all hazardous operations.	All hazardous operations covered by a procedure, posted at the job location, with an annual documented review to determine adequacy.
3. Employee selection and placement.	Only pre-employment physical examination given.	In addition, an aptitude test is administered to new employees.	In addition to "Fair," new employees' past safety record is considered in their employment.	In addition to "Good," when employees are considered for promotion, their safety attitude and record is considered.
4. Emergency and disaster control plans.	No plan or procedures.	Verbal understanding on emergency procedures.	Written plan outlining the minimum requirements.	All types of emergencies covered with written procedures. Responsibilities are defined with backup personnel provisions.
5. Direct management involvement.	No measurable activity.	Follow-up on accident problems.	In addition to "Fair," management reviews all injury and property damage reports and holds supervision accountable for verifying firm corrective measures.	In addition to "Good," reviews all investigation reports. Loss Control problems are treated as other operational problems in staff meeting.
6. Plant safety rules.	No written rules.	Plant safety rules have been developed and posted.	Plant safety rules are incorporated in the plant work rules.	In addition, plant work rules are firmly enforced and updated at least annually.

B. INDUSTRIAL HAZARD CONTROL

Activity	Poor	Fair	Good	Excellent
1. Housekeeping – storage of materials, etc.	Housekeeping is generally poor. Raw materials, items being processed and finished materials are poorly stored.	Housekeeping is fair. Some attempts to adequately store materials are being made.	Housekeeping and storage of materials are orderly. Heavy and bulky objects well stored out of aisles, etc.	Housekeeping and storage of materials are ideally controlled.
2. Machine guarding.	Little attempt is made to control hazardous points on machinery.	Partial, but inadequate or ineffective, attempts at control are in evidence.	There is evidence of control which meets applicable Federal and State requirements, but improvement may still be made.	Machine hazards are effectively controlled to the extent that injury is unlikely. Safety of operator is given prime consideration at time of process design.
3. General area guarding.	Little attempt is made to control such hazards as: unprotected floor openings; slippery or defective floors; stairway surfaces; inadequate illumination, etc.	Partial but inadequate attempts to control these hazards are evidenced.	There is evidence of control which meets applicable Federal and State requirements, but further improvement may still be made.	These hazards are effectively controlled to the extent that injury is unlikely.
4. Maintenance of equipment, guards, handtools, etc.	No systematic program of maintaining guards, handtools, controls and other safety features of equipment, etc.	Partial, but inadequate or ineffective maintenance.	Maintenance program for equipment and safety features is adequate. Electrical handtools are tested and inspected before issuance, and on a routine basis.	In addition to "Good," a preventive maintenance system is programmed for hazardous equipment and devices. Safety reports filed and safety department consulted when abnormal conditions are found.
5. Material handling – hand and mechanized.	Little attempt is made to minimize possibility of injury from the handling of materials.	Partial but inadequate or ineffective attempts at control are in evidence.	Loads are limited as to size and shape for handling by hand, and mechanization is provided for heavy or bulky loads.	In addition to controls for both hand and mechanized handling, adequate measures prevail to prevent conflict between other workers and material being moved.
6. Personal protective equipment – adequacy and use.	Proper equipment not provided or is not adequate for specific hazards.	Partial but inadequate or ineffective provision, distribution and use of personal protective equipment.	Proper equipment is provided. Equipment identified for special hazards, distribution of equipment is controlled by supervisor. Employee is required to use protective equipment.	Equipment provided complies with standards. Close control maintained by supervision. Use of safety equipment recognized as an employment requirement. Injury record bears this out.

C. FIRE CONTROL & INDUSTRIAL HYGIENE

Activity	Poor	Fair	Good	Excellent
1. Chemical hazard control references.	No knowledge or use of reference data.	Data available and used by foremen when needed.	In addition to "Fair," additional standards have been requested when necessary.	Data posted and followed where needed. Additional standards have been promulgated, reviewed with employees involved and posted.
2. Flammable and explosive materials controls.	Storage facilities do not meet fire regulations. Containers do not carry name of contents. Approved dispensing equipment not used. Excessive quantities permitted in manufacturing areas.	Some storage facilities meet minimum fire regulations. Most containers carry name of contents. Some approved dispensing equipment in use.	Storage facilities meet minimum fire regulations. Most containers carry name of contents. Approved equipment generally is used. Supply at work area is limited to one day requirement. Containers are kept in approved storage cabinets.	In addition to "Good," storage facilities exceed the minimum fire regulations and containers are always labeled. A strong policy is in evidence relative to the control of the handling, storage and use of flammable materials.
3. Ventilation – fumes, smoke and dust control.	Ventilation rates are below industrial hygiene standards in areas where there is an industrial hygiene exposure.	Ventilation rates in exposure areas meet minimum standards.	In addition to "Fair," ventilation rates are periodically measured, recorded and maintained at approved levels.	In addition to "Good," equipment is properly selected and maintained close to maximum efficiency.
4. Skin contamination control.	Little attempt at control or elimination of skin irritation exposures.	Partial, but incomplete program for protecting workers. First-aid reports on skin problems are followed up on an individual basis for determination of cause.	The majority of workmen instructed concerning skin-irritating materials. Workmen provided with approved personal protective equipment or devices. Use of this equipment is enforced.	All workmen informed about skin-irritating materials. Workmen in all cases provided with approved personal protective equipment or devices. Use of proper equipment enforced and facilities available for maintenance. Workers are encouraged to wash skin frequently. Injury record indicates good control.
5. Fire control measures.	Do not meet minimum insurance or municipal requirements.	Meets minimum requirements.	In addition to "Fair," additional fire hoses and/or extinguishers are provided. Welding permits issued. Extinguishers on all welding cars.	In addition to "Good," a fire crew is organized and trained in emergency procedures and in the use of fire-fighting equipment.
6. Waste – trash collection and disposal, air/water pollution.	Control measures are inadequate.	Some controls exist for disposal of harmful wastes or trash. Controls exist but are ineffective in methods or procedures of collection and disposal. Further study is necessary.	Most waste disposal problems have been identified and control programs instituted. There is room for further improvement.	Waste disposal hazards are effectively controlled. Air/water pollution potential is minimal.

D. SUPERVISORY PARTICIPATION, MOTIVATION, & TRAINING

Activity	Poor	Fair	Good	Excellent
1. Line supervisor safety training.	All supervisors have not received basic safety training.	All shop supervisors have received some safety training.	All supervisors participate in division safety training session a minimum of twice a year.	In addition, specialized sessions conducted on specific problems.
2. Indoctrination of new employees.	No program covering the health and safety job requirements.	Verbal only.	A written handout to assist in indoctrination.	A formal indoctrination program to orientate new employees is in effect.
3. Job hazard analysis.	No written program.	Job hazard analysis program being implemented on some jobs.	JHA conducted on majority of operations.	In addition, job hazard analyses performed on a regular basis and safety procedures written and posted for all operations.
4. Training for specialized operations (fork trucks, grinding, press brakes, punch presses, solvent handling, etc.)	Inadequate training given for specialized operations.	An occasional training program given for specialized operations.	Safety training is given for all specialized operations on a regular basis and retraining given periodically to review correct procedures.	In addition to "Good," an evaluation is performed annually to determine training needs.
5. Internal self-inspection.	No written program to identify and evaluate hazardous practices and/or conditions.	Plant relies on outside sources, i.e., Insurance Safety Engineer, and assumes each supervisor inspects his area.	A written program outlining inspection guidelines, responsibilities, frequency and follow-up is in effect.	Inspection program is measured by results, i.e., reduction in accidents and costs. Inspection results are followed up by top management.
6. Safety promotion and publicity.	Bulletin boards and posters are considered the primary means for safety promotion.	Additional safety displays, demonstrations, films, are used infrequently.	Safety displays and demonstrations are used on a regular basis.	Special display cabinets, windows, etc., are provided. Displays are used regularly and are keyed to special themes.
7. Employee/ supervisor safety contact and communication.	Little or no attempt made by supervisor to discuss safety with employees.	Infrequent safety discussions between supervisor and employees.	Supervisors regularly cover safety when reviewing work practices with individual employees.	In addition to items covered under "Good," supervisors make good use of the shop safety plan and regularly review job safety requirements with each worker. They contact at least one employee daily to discuss safe job performance.

E. ACCIDENT INVESTIGATION, STATISTICS & REPORTING PROCEDURES

Activity	Poor	Fair	Good	Excellent
1. Accident investigation by line personnel.	No accident investigation made by line supervision.	Line supervision makes investigations of only medical injuries.	Line supervision trained and makes complete and effective investigations of all accidents; the cause is determined; corrective measures initiated immediately with a completion date firmly established.	In addition to items covered under "Good," investigation is made of every accident within 24 hours of occurrence. Reports are reviewed by the department manager and plant manager.
2. Accident cause and injury location analysis and statistics.	No analysis of disabling and medical cases to identify prevalent causes of accidents and location where they occur.	Effective analysis by both cause and location maintained on medical and first-aid cases.	In addition to effective accident analysis, results are used to pinpoint accident causes so accident prevention objectives can be established.	Accident causes and injuries are graphically illustrated to develop the trends and evaluate performance. Management is kept informed on status.
3. Investigation of property damage.	No program.	Verbal requirement or general practice to inquire about property damage accidents.	Written requirement that all property damage accidents of $50 and more will be investigated.	In addition, management requires a vigorous investigation effort on all property damage accidents.
4. Proper reporting of accidents and contact with carrier.	Accident reporting procedures are inadequate.	Accidents are correctly reported on a timely basis.	In addition to "Fair," accident records are maintained for analysis purposes.	In addition to "Good," there is a close liaison with the insurance carrier.

RATING FORM

	Poor	Fair	Good	Excellent	Comments
A. ORGANIZATION & ADMINISTRATION					
1. Statement of policy, responsibilities assigned.	0	5	15	20	
2. Safe operating procedures (SOPs).	0	2	15	17	
3. Employee selection and placement.	0	2	10	12	
4. Emergency and disaster control plans.	0	5	15	18	
5. Direct management involvement.	0	10	20	25	
6. Plant safety rules.	0	2	5	8	
Total value of circled numbers	______	+ ______	+ _____	+ _____	x .20 Rating ______
B. INDUSTRIAL HAZARD CONTROL					
1. Housekeeping – storage of materials, etc.	0	4	8	10	
2. Machine guarding.	0	5	16	20	
3. General area guarding.	0	5	16	20	
4. Maintenance of equipment guards, handtools, etc.	0	5	16	20	
5. Material handling – hand and mechanized.	0	3	8	10	
6. Personal protective equipment – adequacy and use.	0	4	16	20	
Total value of circled numbers	______	+ ______	+ _____	+ _____	x .20 Rating ______
C. FIRE CONTROL & INDUSTRIAL HYGIENE					
1. Chemical hazard control references.	0	6	17	20	
2. Flammable and explosive materials control.	0	6	17	20	
3. Ventilation – fumes, smoke and dust control.	0	2	8	10	
4. Skin contamination control.	0	3	10	15	
5. Fire control measures.	0	2	8	10	
6. Waste – trash collection and disposal, air/water pollution.	0	7	20	25	
Total value of circled numbers	______	+ ______	+ _____	+ _____	x .20 Rating ______

RATING FORM

	Poor	Fair	Good	Excellent	Comments
D. SUPERVISORY PARTICIPATION, MOTIVATION, & TRAINING					
1. Line supervisor safety training.	0	10	22	25	
2. Indoctrination of new employees.	0	1	5	10	
3. Job hazard analysis.	0	2	8	10	
4. Training for specialized operations.	0	2	7	10	
5. Internal self-inspection.	0	5	14	15	
6. Safety promotion and publicity.	0	1	4	5	
7. Employee/supervisor contact and communication.	0	5	20	25	
Total value of circled numbers	______	+ ______	+ _____	+ _____	x .20 Rating ______

	Poor	Fair	Good	Excellent	Comments
E. ACCIDENT INVESTIGATION, STATISTICS & REPORTING PROCEDURES					
1. Accident investigation by line supervisor.	0	10	32	40	
2. Accident cause and injury location analysis and statistics.	0	3	8	10	
3. Investigation of property damage.	0	10	32	40	
4. Proper reporting of accidents and contact with carrier.	0	3	8	10	
Total value of circled numbers	______	+ ______	+ _____	+ _____	x .20 Rating ______

SUMMARY

The numerical values below are the weighted ratings calculated on the rating sheets. The total becomes the overall score for the location.

A. Organization & Administration _________

B. Industrial Hazard Control _________

C. Fire Control & Industrial Hygiene _________

D. Supervisory Participation, Motivation, & Training _________

E. Accident Investigation, Statistics & Reporting Procedures _________

TOTAL REPORTING _________

Appendix B

A second example of an excellent safety approach follows. It is reprinted here with permission from MeadWestvaco Corporation.

MEAD MANAGEMENT SYSTEM: "GUIDELINES FOR SAFETY EXCELLENCE"

Overview

The Mead "Management System Guidelines For Safety Excellence" ("Guidelines For Safety Excellence") were developed by the Safety Excellence Task Force (SETF). These guidelines were offered to all Mead locations as an example of a comprehensive management system designed to focus on the prevention of occupational injuries and illnesses. Considerable research, including benchmark learning, went into the development of the guidelines. The content and use of the guidelines have been endorsed by the SETF and Leadership Council as a framework to achieve safety excellence in Mead.

A safety management system is an integrated, documented process that outlines how an organization intends to develop and achieve its injury and illness prevention objectives. Mead has captured the best management practices in the nine (9) elements listed in the Safety Excellence Guidelines. The guidelines include a brief synopsis of each element followed by suggested Objectives and Indicators/Measures. The Objectives describe how an organization can meet the subject element provisions whereas, the Indicators/Measures provide a description of activities that can be used to measure an organization's success in implementing each objective for that element.

The Guidelines For Safety Excellence, when used as a reference during an annual self-assessment, allow each location to compare themselves to best management practices that have resulted in safety excellence.

Using the "Guidelines For Safety Excellence"

These guidelines should be used as a reference during all annual self-assessments. The guidelines offer specific activities that, when implemented, will support the location's efforts toward achieving safety excellence. With that in mind, the following steps should be considered when using these guidelines:

Step 1: Ensure that all self-assessment team members have been educated in Mead's Safety Excellence Initiative, these Management System Guidelines For Safety Excellence and the companion Safety Excellence Self-Assessment Tool.

Step 2: Conduct an initial assessment of your management system for injury and illness prevention using the Self-Assessment Tool.

Step 3: Compare the results of the self-assessment to the "best practices" outlined in these guidelines and identify opportunities for improvement in the management system. Note: when used in conjunction with the Petersen Safety Perception Survey, the same procedures can be followed.

Step 4: Develop an action plan outlining specific action items needed to improve safety performance.

Step 5: Integrate the "Action Plan" into the location's strategic business plan.

Step 6: Reassess the management system on a periodic basis to identify progress made as well as identify opportunities for further improvement.

ELEMENT I: MANAGEMENT COMMITMENT AND LEADERSHIP

Safety Excellence begins with the commitment and leadership of senior management. This commitment must be demonstrated through visible leadership and participation in daily safety activities and encouragement of employee participation in safety efforts. Senior management regards the safety of employees as a fundamental value of the organization and applies its commitment to safety equally with other organizational drivers.

OBJECTIVES	INDICATORS/MEASURES
a. Establish and communicate a policy statement to set clear expectations regarding safe and healthful work practices and conditions for each facility	• Written, up-to-date policy is established and communicated to all employees • All employees at the facility are aware of the policy, its contents, and understand the expectations • Active participation of employees is encouraged to improve safety performance
b. Operate consistently with a philosophy that all occupational injuries and illness are preventable	• Involvement of all employees is actively pursued including wide communication of issues with employees • Systems are in place to encourage reporting of hazards by employees • Employees are included in the process of identifying hazards as well as the identification of solutions to eliminate or control the hazards • Safe behaviors are recognized, acknowledged, and praised (positive reinforcement) at all levels • System of intervention into unsafe behavior is in place at all levels
c. Demonstrated management involvement and support in the safety and health process	• Safety issues are included on meeting agendas • Management is visible to employees by spending time on the operating floor • Management sets a positive example in daily work activities • Managers are as conversant regarding safety as they are with other business issues

	• Safety Action Plan established and communicated to the workforce • Clear understanding that safety is the responsibility of operations management
d. Establish goals and objectives that are aligned throughout the organization to drive step change improvements in safety performance	• Self-assessments using the Safety Excellence Assessment Tool are conducted on a periodic basis to identify opportunities for system improvements • Findings from the self-assessment are validated on a periodic basis using the Mead Perception Survey • Written action plans are developed for the facility, based on findings from the self-assessment • Annual performance objectives are established, linked to action plans, and tracked • Goals and objectives are developed with input by all employees and communicated to all stakeholders • All employees are knowledgeable and understand the goals and objectives for performance of assigned responsibilities • Safety Action Plan is integrated into the business plan of the facility
e. Allocate adequate resources to support achievement of the objectives defined in the Safety Action Plan	• Adequate staffing exists to support line management's implementation of the safety action plan and safety system needs • Sufficient time is allotted for line management to execute safety responsibilities • A funding mechanism exists to implement safety/health improvement plans • All individuals are qualified and adequately trained and equipped to successfully perform their safety and health roles and responsibilities

ELEMENT II: ROLES, RESPONSIBILITIES, AND ACCOUNTABILITIES

Effective implementation of a safety management system is a line management responsibility with active participation of employees at all levels of the organization. Roles and responsibilities for safety and health are clearly defined and there is a strong accountability structure.

OBJECTIVES	INDICATORS/MEASURES
a. Roles and responsibilities for safety and health are clearly defined and communicated for all positions in the organization	• Up-to-date roles and responsibilities are clearly defined and communicated for all positions (e.g., job descriptions, AMCs/AMSs) • All employees at the facility are aware of and understand the safety and health roles and responsibilities for their position • Defined activities are based on the level in the organization and are aligned with the organization's goals and objectives • Behavior of all employees demonstrates commitment to individual roles and responsibilities
b. Establish a system to hold employees at all levels of the organization accountable for performance of assigned responsibilities	• Performance measurement system is clearly understood by employees at all levels of the organization • Performance measures are objective and quantifiable • Specific accident prevention activities are incorporated into individual accountability contracts (e.g., AMC, Accountability Management Contracts; AMS, Accountability Management System) • Performance appraisal process includes evaluation of accident prevention activities • Procedures in place to reward/recognize performance to reinforce the importance of safety and health objectives

- Middle management is held accountable for safety administration within their department and audits the quality of the accident prevention activities delegated to line managers
- Periodic (e.g., quarterly) reviews and assessments are conducted to measure performance against stated objectives and to ensure satisfactory results

c. Employees at all levels of the organization are provided with training to develop and/or enhance skills necessary to fulfill safety roles and responsibilities

- Skills required to drive safety excellence are defined and understood by management
- Assessment of skills required to drive safety excellence is complete
- Assessment of knowledge and understanding of accident prevention skill complete
- Training plan developed and implemented to close skill gaps

ELEMENT III: EMPLOYEE PARTICIPATION

An effective safety management system requires participation of employees at all levels of the organization. Employees are encouraged to provide meaningful input into the operation of the program and the decisions that affect their safety and health.

OBJECTIVES

a. Management provides for and encourages meaningful and significant employee involvement in the safety and health process

INDICATORS/MEASURES

- Employees actively participate in accident prevention activities and accept personal responsibility for working safely

- Employees are involved in accident prevention activities/tasks such as:
  - Inspecting for hazards and unsafe behaviors and recommending corrective actions or controls
  - Analyzing jobs to identify potential hazards and develop safe work procedures
  - Training newly hired employees in safe work practices
  - Assisting in incident investigations
  - Review of new processes and equipment
- Employees participate on safety teams such as safety boards, task teams, focus groups, and ergonomics teams
- Employees are involved in the process of establishing goals and objectives for safety and health

b. Supervisors and managers are held accountable for motivating employees to participate in the safety process

- Safety teams have voluntary employee participation
- Management periodically solicits the workforce inviting participation on safety teams and requesting feedback and input on safety systems

ELEMENT IV: COMMUNICATION

Communication is an integral element of an effective safety management system. Focus is dependent upon free-flowing, two-way communication between management and employees. Management communicates expectations, provides general information regarding safety, and provides guidance to employees. Employees communicate to raise issues for review, make suggestions, and provide feedback regarding clarity of management's communication. This element of a safety management system must establish how management will communicate, how employee input will be obtained, and how the communication system will be enhanced to maximize the value and timeliness of information.

OBJECTIVES	INDICATORS/MEASURES
a. Establish methods for communicating between all levels and functions of the organization	• Systems are established for employee reporting of safety issues and concerns without fear of reprisal and there is a process for management response and feedback • Methods established for communicating goals and objectives, safety performance (trailing indicators), accident prevention activities (leading indicators), opportunities for improvement, successes and recognition and other safety/health information • Safe work practices (e.g., safety rules) are consistently communicated to all employees and are actively monitored

ELEMENT V: BEHAVIORAL SKILLS

Since the root cause of most accidents is related to behavior, successful safety programs focus on human behavior. The most direct influence on an employee's behavior is the positive or negative consequence of their behavior. This element of a safety management system must describe how employees will be educated to recognize both safe and unsafe behavior; how behavioral observations will be made; how safe behavior will be encouraged, reinforced, and measured; and how intervention will occur to address unsafe behavior.

OBJECTIVES	INDICATORS/MEASURES
a. Establish a system to promote safe behaviors throughout all levels of the organization	• Provide training to all levels in the organization on the role of behaviors in achieving safety excellence • Safe behaviors are recognized and rewarded

	• Supervisors have become more effective coaches based on behavioral knowledge • Routine behavioral observations of workplace activity are conducted by all levels of the organization • Managers provide leadership by actively participating in behavioral observations • Observations are being completed and tracked • Data gathered during the observations are used to identify underlying root causes
b. Create an atmosphere where interventions into at-risk behaviors are encouraged	• Interventions occur horizontally and vertically throughout the organization • Individuals give and receive positive and negative feedback equally well

ELEMENT VI: HAZARD RECOGNITION AND CONTROL

The elimination or control of hazards in the workplace requires a proactive system to identify, assess, and manage situations that have the potential to result in personal injury, illness, or property damage. This system must be applied to both new and existing facilities and procedures. To achieve safety excellence, this element of the safety management system must define the types of review that will be completed, and the method for eliminating or controlling the identified hazards.

OBJECTIVES	INDICATORS/MEASURES
a. Establish design/modification reviews that consider safety and health	• Procedure is in place to review the safety and health impact of new installations or process changes at the design stage • There is a multi-functional participation in the review process including hourly employees

b. Establish a system to routinely identify and correct physical hazards in the workplace	• Findings and recommendations are documented • Needed corrective actions are communicated with employees • System established to ensure timely correction • Follow-up inspections are made to ensure corrective action has been implemented and that there is no recurrence of the identified hazard or behavior
c. An effective occupational health program has been established to recognize, evaluate, and control employee exposure to harmful chemical, physical, and biological agents, and associated occupational illnesses	• The facility uses trained occupational health professionals (i.e., industrial hygienists, ergonomists, radiation safety experts) to systematically evaluate employee exposures and control strategies • A validated qualitative exposure assessment system is used to define employee monitoring strategies • Employee exposure monitoring and qualitative evaluation of results are conducted using sound statistical methodology • Employees are informed of their exposures to chemical, physical, and biological agents and have access to occupational health professionals who can explain the results • Exposures to chemical, physical, and biological agents in excess of current recommended values are given a high priority for corrective action using a hierarchy of engineering controls, administrative controls, and personal protective equipment • Medical professionals review exposure assessment results and employee medical data to develop and implement medical surveillance programs (where necessary)

ELEMENT VII: EDUCATION AND TRAINING

Achieving safety excellence is dependent upon having the right personnel in place with the necessary skills and knowledge required to properly perform assigned roles and responsibilities. Training is necessary to implement management's commitment to accident prevention and to ensure that supervisors and employees understand the policies, procedures, and safe work practices to prevent exposure to hazards.

OBJECTIVES	INDICATORS/MEASURES
a. Identify training needs for all levels of the organization to ensure necessary skills are adequate to fulfill safety roles and responsibilities as well as perform work safely	• Education and training plan developed to address safety/health training needs for employees at each level in the organization • Assessment of skills necessary to perform safety roles and responsibilities, work safely, and sustain regulatory compliance is complete • Appropriate training completed for those involved in safety and health activities (e.g., behavioral modification, job hazard analysis, coaching and intervention methods, hazard recognition, etc.) in addition to training required for regulatory compliance • Training completed on the facility safety and health policy and safe operating procedures
b. Responsibility assigned to monitor and evaluate training program effectiveness	• Training is documented and documentation is easily retrievable • Training tracking system used to ensure all employees receive required training • Competency in training and subject is verified through observation, skills demonstration, and/or testing • Training plan is periodically evaluated to ensure all required training is scheduled and completed

c. All employees are aware of the hazards to which they may be exposed and methods to prevent or minimize exposure to the hazards, as well as facility policy and regulatory compliance requirements	• Supervisors understand the hazards associated with jobs in their areas, their potential effects on employees, and the methods to eliminate, reduce, or control exposure to the hazards • All employees are made aware of hazards, safe work practices, and procedures associated with their job • Employees demonstrate safe work behaviors and knowledge of facility policies, job hazards, and regulatory compliance requirements

ELEMENT VIII: INCIDENT INVESTIGATION AND ANALYSIS

Effective investigation of incidents is an integral part of successful safety systems. Investigation of incidents should be directed at identifying the underlying factors (root causes) that led to the incident and the identification of corrective measures to prevent recurrence. Findings and recommendations should be shared with affected employees to maximize learning.

OBJECTIVES	INDICATORS/MEASURES
a. Documented systematic method of investigating all incidents emphasizing the importance of identifying root cause(s) and sharing of investigation findings and recommendations	• Procedures for incident investigation are established and understood by all employees • Investigations are objective fact-finding missions, not designed to place blame • Investigations are completed in a timely manner • All OSHA recordable, first-aid and near-miss incidents are investigated as well as spills, releases, vehicle accidents and incidents involving processes regulated by the OSHA Process Safety Management Standard

	• Findings and recommendations of all investigations are documented
b. Implement actions to ensure the quality of incident investigations and reporting	• Where necessary, investigation teams are established to conduct investigations of serious incidents • Training is provided to all employees involved in investigating incidents • Incident reports are routinely evaluated to ensure quality prior to issuance • Root causes are identified and corrective actions address root causes • Investigation reports are reviewed by several levels of the organization and have multiple sign-offs
c. Establish corrective action plans following incident investigations that include review/modification of policies, procedures, process designs and/or training materials	• Corrective actions are directed at root causes, not effects (or results) of the incident • Procedures in place to ensure completion of corrective actions • Individual performance accountability contracts include a measurement for follow-up on closure of recommended actions
d. Feedback is provided to the work force on incident investigation findings and actions taken to prevent recurrence	• Feedback provided to employees via safety meetings, newsletter, bulletin board posting, or other documented means
e. Management encourages the reporting of near-miss and first-aid incidents	• Employees are educated in the importance of reporting near-miss incidents and first-aid incidents • Near-miss and minor injury incidents are reported • Feedback mechanisms exist for communicating corrective measures on near-miss incidents

ELEMENT IX: PERFORMANCE MEASUREMENT

Performance measurement is integral to achieving safety excellence. The safety management system must be continuously evaluated to ensure the system is effective. This element of a safety management system must describe how performance will be measured and the subsequent actions that will be taken to improve the system.

OBJECTIVES	INDICATORS/MEASURES
a. Select relevant methods to measure safety activities (e.g., leading indicators) and safety performance (e.g., trailing indicators)	• Utilize the Safety Excellence Self-Assessment Tool to identify opportunities for improving the safety management system • Findings from the self-assessment are validated on a periodic basis using the Petersen Safety Perception Survey • Based on findings from the self-evaluation and/or perception survey, written action plans are developed, implemented, and monitored to ensure completion and measure effectiveness
b. Safety Action Plan is measured, tracked, and assessed on a periodic basis, with results being posted in the workplace	• Safety Action Plan is periodically reviewed to determine implementation status and results • Review process eliminates ineffective safety activities • Periodic reviews consider additional activities necessary to achieve safety excellence

Appendix C

Statistical process control (SPC) concepts were not really meant to be measures; they were meant to be problem-solving tools. After data was collected, SPC would help determine what problems created this data.

SPC is not a measurement tool as much as a batch of analytical tools. However, it has been included in this Appendix because it provides valuable information that the safety process often overlooks. Also, SPC is included because (1) it can lead to many solutions not reached through traditional safety methodology, and (2) it can help to integrate safety into the mainstream of management.

SPC is one part of the total quality improvement process, or the continuous improvement process, which are both parts of the Deming philosophy. SPC addresses:

1. The analysis of the data.
2. The determination of why the data came out the way it did.
3. What to do about control for the future.

The concept of continuous improvement has been with us for many years. Businesses have used the concept in improving quality and productivity, but seldom in improving safety, yet the same approach can apply to any aspect of management.

Continuous improvement comes under many labels: total quality management, statistical process control, the Deming philosophy, and Six Sigma are a few. All of these concentrate on gradual, ongoing improvement of the process, whether it is quality, productivity, or safety. The concepts are proven; they work in other areas, and they will work in safety as well.

First, we'll look at the Deming philosophies and attempt to adapt them to safety and then move on to the SPC tools, which are very useable for us in safety.

Safety and total quality management (TQM) work hand-in-hand, unless safety practitioners choose to block the union because they perceive it to be different and threatening. Some within the profession contend that the goal of a safety practitioner is to work him or her out of a job. Integrating safety into TQM may be the best way to do just that.

The real question, however, is: Do safety practitioners want this marriage? Consider these potential consequences:

- Replacement of job safety analyses (JSAs) with flow diagrams, which will diagnose system weaknesses. Such weaknesses could be diagnosed by those who work most closely with the system elements—the employees.
- Replacement of accident investigation procedures with fishbone diagrams to search for multiple causes.
- Use of behavioral sampling and perception surveys to measure effectiveness and identify weaknesses.
- Management acceptance and understanding of the safety management process.
- A new organizational culture, with safety perceived as a high corporate value.
- Accountability.

In most companies, quality of performance means, initially, quality of product and increased productivity. Some companies include quality of safety performance at a much later point, often as an afterthought.

Employee involvement is central to the TQM philosophy, which also includes these concepts:

- Building a new organizational culture that embraces safety
- Using new tools to solve problems
- Continuous improvement of the process
- Using upstream measures to monitor progress.

Perhaps the best description of TQM is captured in Deming's fourteen "Obligations of Management" presented in Exhibit C-1.

■ Exhibit C-1 **Dr. W. Edwards Deming's 14 Obligations of Top Management**

1. Innovate and allocate resources to fulfill the long-range needs of the company and customer rather than short-term profitability.
2. Discard the old philosophy of accepting defective products.
3. Eliminate dependence on mass inspection for quality control; instead, depend on process control through statistical techniques.
4. Reduce the number of multiple source suppliers. Price has no meaning without an integral consideration for quality. Encourage suppliers to use statistical process control.
5. Use statistical techniques to identify the two sources of waste—system (85%) and local faults (15%); strive to constantly reduce this waste.
6. Institute more thorough, better job-related training.
7. Provide supervision with knowledge of statistical methods; encourage use of these methods to identify which defects should be investigated for solution.
8. Reduce fear through the organization by encouraging open, two-way, nonpunitive communication. The economic loss resulting from fear to ask questions or report trouble is appalling.
9. Help reduce waste by encouraging design, research, and salespeople to learn more about the problems of production.
10. Eliminate the use of goals and slogans to encourage productivity, unless training and management support is also provided.
11. Closely examine the impact of work standards. Do they consider quality or help anyone do a better job?
12. Institute rudimentary statistical training on a broad scale.
13. Institute a vigorous program for retraining people in new skills, to keep up with changes in materials, methods, product designs, and machinery.
14. Make maximum use of statistical knowledge and talent in your company.

If Deming's "Obligations of Management" were rewritten in safety jargon, they might read as follows:

Management Safety Obligations

1. Concentrate on the long-range goal of developing a world-class system, not on short-term, annual accident goals.
2. Discard the philosophy of accepting accidents; they are unacceptable.

3. Use statistical techniques to identify the two sources of accidents: the system and human error.

4. Institute more thorough job-skills training.

5. Eliminate dependence on accident investigation. Instead, use proactive approaches such as behavioral sampling, fishbone diagrams, flowcharts, and others, to reveal system flaws and achieve continuous system improvement.

6. Provide supervisors (and employees) with knowledge of statistical methods (sampling, control charts, etc.) and ensure that these tools are used to identify areas needing additional study.

7. Reduce fear throughout the organization by encouraging all employees to report system defects and help find solutions.

8. Reduce accidents by designing safety into the process. Train research and design personnel in safety concepts.

9. Eliminate the use of slogans, incentives, posters and gimmicks to encourage safety.

10. Examine work standards to remove accident traps, unless training and management support is also provided.

Other aspects of TQM are valuable as well. In fact, items such as the following would be necessary in safety:

- Asking employees to define and solve company problems and identify system weaknesses.
- Providing employees with simple tools to solve problems. These include Pareto charts to determine problems, fishbone diagrams to help brainstorm problem causes, flowcharts to observe the system, and scatter diagrams to determine correlations.
- Replacing accident-based statistics with other upstream measures (i.e., behavioral sampling, etc.).

- Replacing accident-based statistics with alternative downstream measures (i.e., employee perception surveys, employee interviews).

The ten management safety obligations listed earlier represent a marked departure from traditional safety beliefs. Under these new corporate obligations:

1. Progress is not measured by accident rates.
2. Safety becomes a system, rather than a program.
3. Statistical techniques drive continuous improvement efforts in safety.
4. Accidents are caused by a defective management system and a weak safety culture.
5. Many methods can be used to shape behavior, not merely the three Es (engineering, education, and enforcement).
6. No magic pill can be prescribed. Practitioners must determine which approaches will work best, depending on situational demands.
7. Executives must provide safety leadership.
8. Decisions made at the bottom, by affected employees, are most effective.

The move toward TQM in safety means refuting many traditional concepts and beliefs such as:

1. Irresponsible acts and conditions cause accidents.
2. The three Es of safety are essential to safety programs.
3. Law compliance is sufficient.
4. The executive role is only to sign policy.
5. Management creates safety rules; employees follow them.

These beliefs should be replaced with the following axioms:

- Accident investigations are either reformed or eliminated.
- Safety sampling and statistical process-control tools are used.
- Blame for "unsafe acts" is completely eliminated.
- Focus is on improving the safety system.
- "Whistle blowers" are encouraged and supported.
- Employee involvement in problem solving and decision making is formalized via corporate procedures.
- Ergonomic solutions are designed into the workplace.
- Safety slogans and gimmicks are eliminated.
- Emphasis is placed on removing system traps that cause human error.

The basic TQM concepts are:

- **Employee involvement.** Decide what employee involvement means. Does it mean: Asking for input *before* management decisions are finalized? Sharing the decision-making process? Allowing employees to make decisions? Once the level of involvement is defined, take the short-term steps needed to move in that direction. These include confirming that management is credible and has done everything possible to ensure safety.
- **New culture.** Safety must be perceived as a key value. Again, ensure that management is credible. Determine the status of the company's safety culture and take steps to establish a new culture.
- **New tools.** Train employees to use problem-solving tools. Create a structure in which they can effectively use these tools.
- **Continuous improvement and use of the best activity and results measures.** These concepts require that the influence of accident-based rates be *completely* removed from upstream measures (and likely from downstream measures as well). Rates must be replaced by behavioral sampling, perception surveys, and other tools that have statistical reliability.

Managing With Data

Data serves as the basis for appropriate action. Data provides the driving reason for making decisions. Most data can be described using the following two broad filters:

- Results measures tell us if we have achieved a goal or objective.
- Activity measures are interim indicators that tell us how well we are progressing toward reaching a goal or objective.

Statistical safety control (SSC) is the utilization of some common statistical process control concepts in safety. The SPC tools that can be used most easily are the Pareto chart, the fishbone diagram, the flowchart, and the control chart.

SPC is a popular concept in the United States today, although this hasn't always been the case. American industry basically spurned the whole idea of SPC in the 1950s; now it's endorsing it.

The United States ignored SPC and found its productivity dropped 15 percent between 1960 and 1980. The Japanese bought it and increased their productivity by 150 percent in the same period. And their SPC thrusts were for quality, not productivity.

In the 1980s, American managers rushed to Japan to learn their secret, only to find it was what the United States had rejected twenty years earlier.

The use of statistical methods is not an all-purpose remedy for every corporate problem. But it is a rational, logical, and organized way to create a system that can assure ongoing improvements in quality and productivity simultaneously. This is also true for safety.

The SPC method is really a two-pronged approach in increasing employee involvement, using some SPC tools to solve problems. Using SPC in safety allows specific problems to be attacked and solved while the system is being monitored. The SPC tools can be defined as follows:

Tools to Describe the Data

1. Line graph (Exhibit C-2)
2. Bar charts (Exhibit C-3)
3. Paretos (Exhibit C-4)
4. Control charts (Exhibit C-5 and C-5a–f)
5. Fishbone diagrams (Exhibit C-6)

Tools for Decision Making

6. Cost benefit analyses (Exhibit C-7)
7. Failure mode and effect analysis (FMEA) (Exhibits C-8a-c)

Tools to Determine Cause

8. Flowcharts (Exhibit C-9)

Tools to Determine Relationships

9. Scatter diagrams (Exhibit C-10)

1. LINE GRAPH

The line graph (Exhibit C-2) is a line of collected dots used to show data variation over time. It is the best way to display changes over time. It helps to interpret the data being collected and is a means of communicating changes graphically to others. It also is the most frequently used of all graphs because it is simple to make and gives a clear picture of change. The line graph:

- Quickly gives a picture of data.
- Is easy to create.
- Identifies areas of process variation.
- Displays the current condition.
- Summarizes the results of an implemented solution.

■ Exhibit C-2 **Sample line graph**

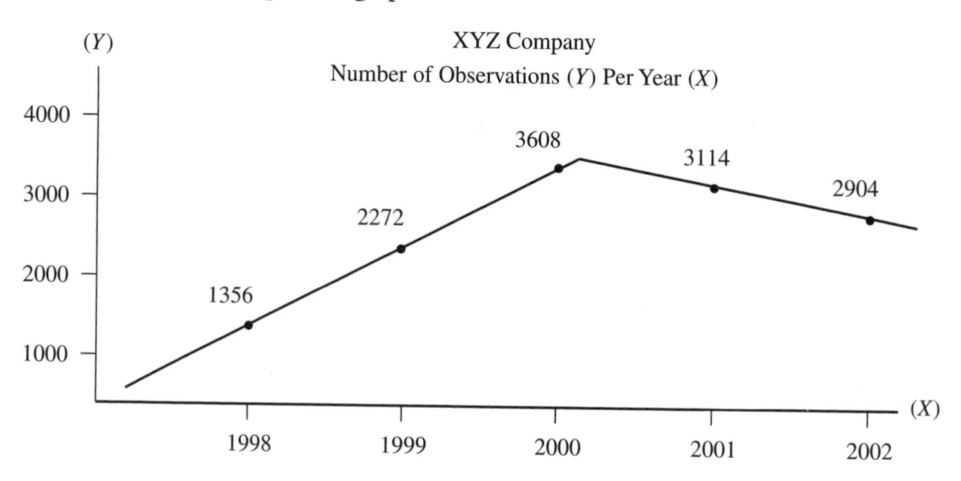

2. BAR CHART

Exhibit C-3 shows a series of bars representing the sizes of different, but related items. It helps to compare categories of data to one another and is a tool for breaking data into its component parts. It can be used to compare data from two or more similar items.

■ Exhibit C-3 Basic bar chart

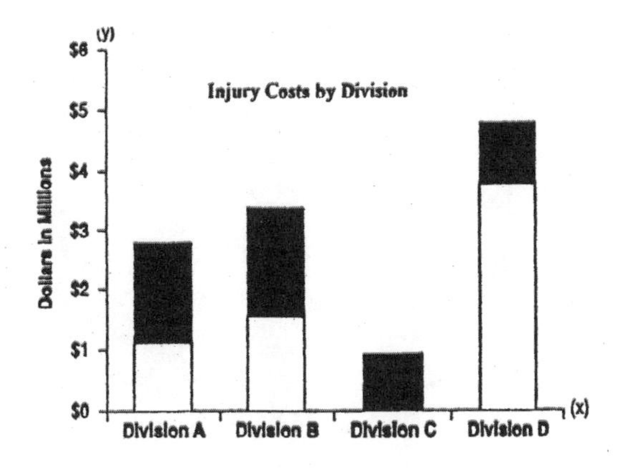

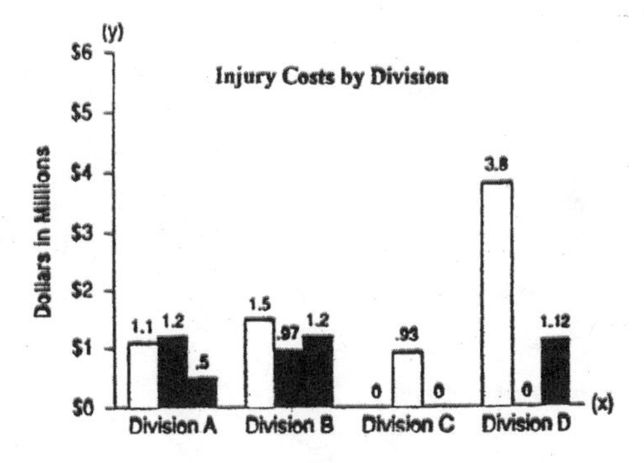

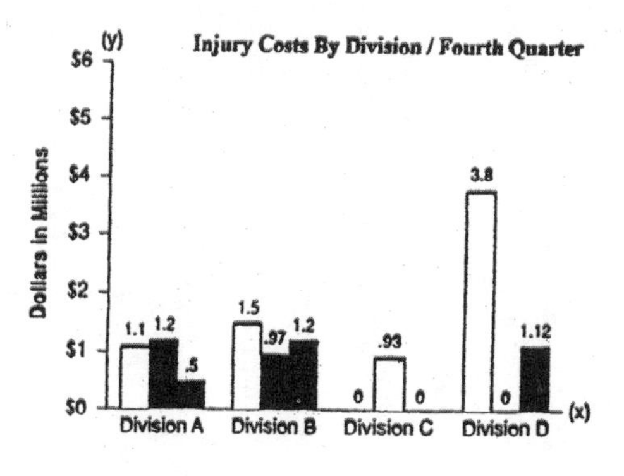

3. PARETO CHART

The Pareto is a vertical bar graph that arranges events according to frequency of occurrence. It helps determine which problems to solve in what order. It directs attention to what's most important, and it focuses attention on the vital few, not the trivial many. A Pareto chart:

- Directs attention and focuses effort.
- Shows where to concentrate limited resources first to make the greatest improvement.
- Tracks progress and documents improvement by comparing changes from one time period to another.
- Is especially valuable in arranging items like errors, defects, or failures. These are called *zero-based* items because you hope to eliminate them by reducing their number to zero, or as close to zero as possible.
- Separates items into categories, making it easier to identify which particular errors, defects, or failures are the most frequent and most significant.

The chart in Exhibit C-4 was developed by a group of hourly workers to identify why (combinations of reasons) they get soft-tissue injuries in wrists, shoulders, and backs.

Paretos can be used in many ways in safety to plot types of injuries that have occurred. Using the charts in Exhibit C-4, it was decided that a step change could be made by reducing the soft-tissue injuries in two departments: the packers in the packing department and the box handlers in shipping.

■ Exhibit C-4 **Basic Pareto chart of identified injuries shown in descending order of magnitude or frequency by department**

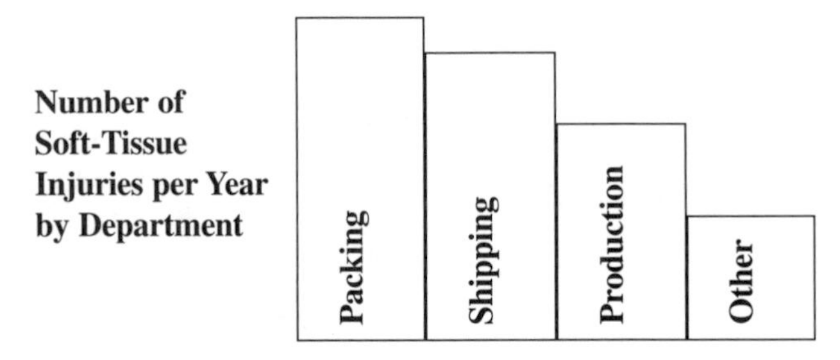

■ Exhibit C-5 Process control chart

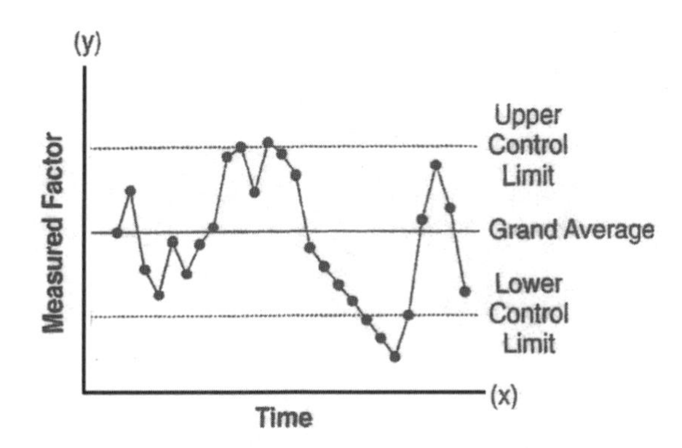

4. CONTROL CHART

This is a type of trend chart (Exhibit C-5) with limits specifically used to track an ongoing process over time to see if it is within limits or control. It shows process variability—whether the process is within limits or in control. The control chart:

- Monitors the results of a process over a given period of time.
- Identifies points that fall outside the control limits and need investigation.
- Helps identify times when the points within the limit lines fall into patterns, also indicating the possibility of a problem.

INTERPRETING CONTROL CHARTS

■ Exhibit C-5a Pattern 1: A point above the upper control limit or below the lower control limit

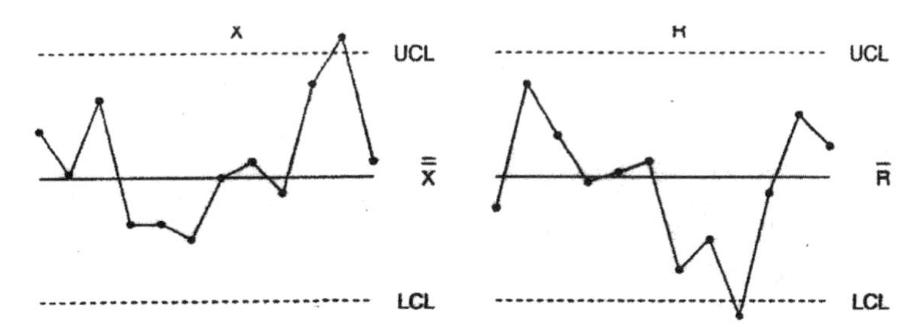

■ Exhibit C-5b **Pattern 2: Trend of seven points in a row increasing or decreasing**

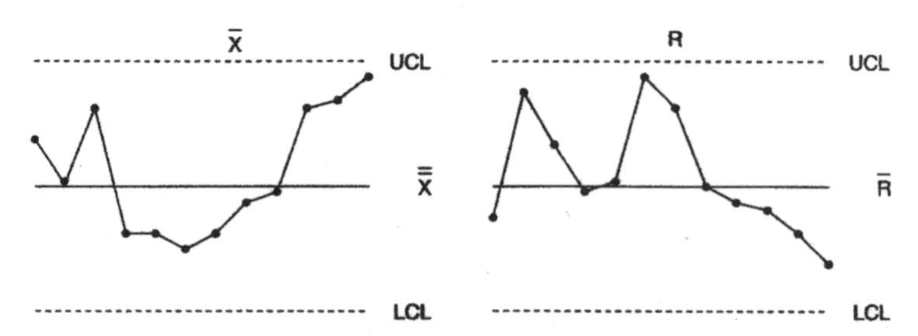

■ Exhibit C-5c **Pattern 3: Cycles or recurring patterns**

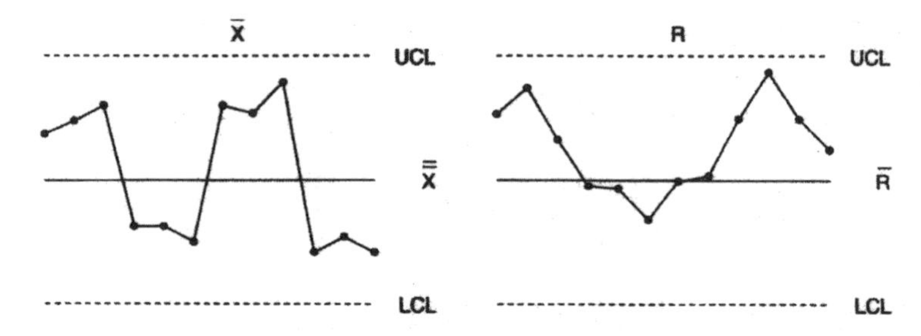

■ Exhibit C-5d **Pattern 4: Seven points in a row on the same side of the centerline**

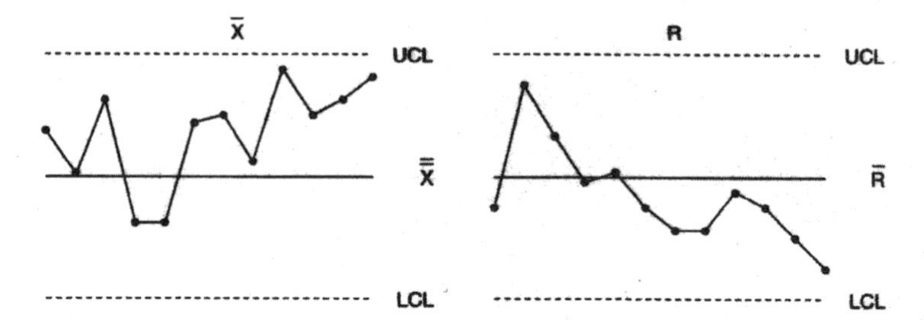

■ Exhibit C-5e **Pattern 5: Two points in a row very close to the upper and lower control limit**

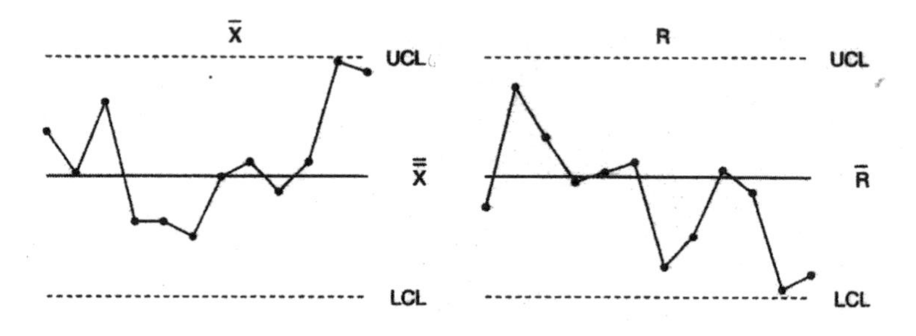

■ Exhibit C-5f **Pattern 6: Ten of eleven points on the same side of the centerline**

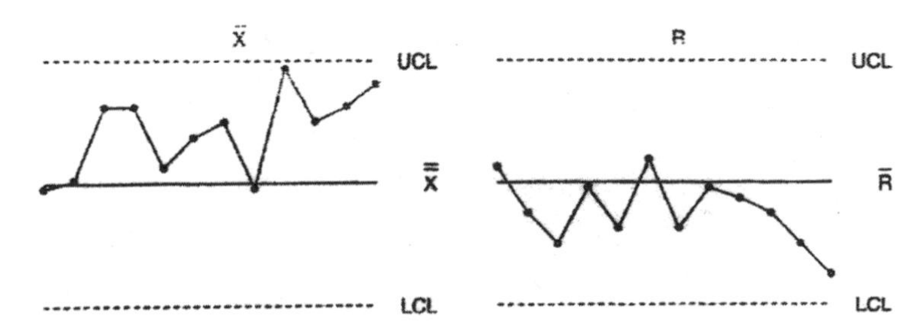

5. CAUSE AND EFFECT DIAGRAM (FISHBONE)

This is a tool for gathering and organizing ideas on possible causes that might lead to a particular effect. It is called a fishbone diagram because it resembles a fish skeleton. It clearly illustrates potential causes of a specific effect. It identifies major causes, examining the methods, materials, people, environment, and equipment. It begins with experienced-based guesses and progresses toward data-based analysis. The fishbone (Exhibit C-6) helps identify the most likely causes of a problem, identifies possible causes for further analysis, and discovers deviations from the norm or patterns.

■ Exhibit C-6 **Fishbone chart of conditions contributing to soft-tissue injuries in a packing or shipping department.**

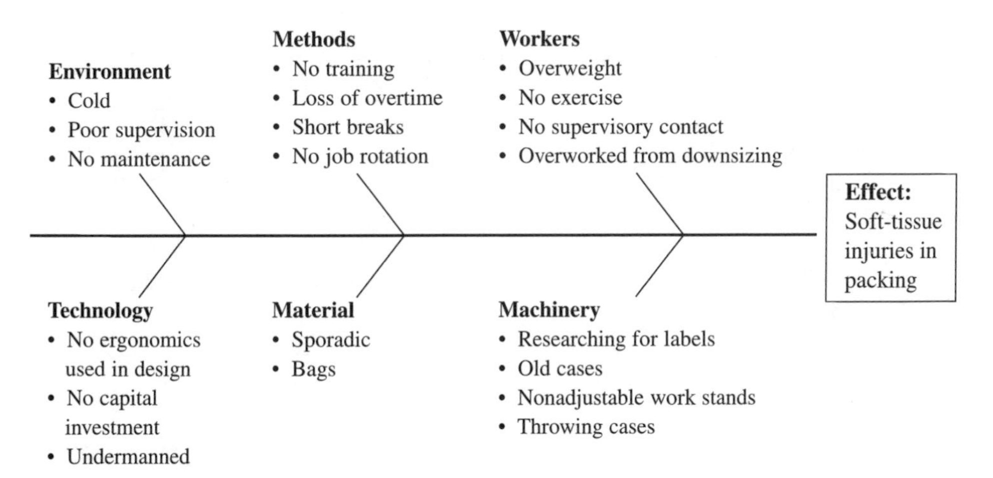

6. COST-BENEFIT ANALYSIS

This is a comparison of the costs and benefits associated with each option under consideration (Exhibit C-7). It should be used to determine the best alternative solution prior to implementation, and be used after implementation as part of the project evaluation to verify costs and benefits.

■ Exhibit C-7 Sample cost-benefit analysis

COST-BENEFIT ANALYSIS
Getting New Equipment to Reduce Injury

COSTS:	
Machine	$ 1,400.00
Installation	300.00
Retraining operator	250.00
Production loss while in transition	500.00
TOTAL COSTS	$ 2,450.00

BENEFITS:	Year 1	Year 2	Year 3	Year 4	
Reduce injuries by 15%	$ 800.00	$ 800.00	$ 850.00	$ 850.00	
Reduce maintenance	500.00	500.00	450.00	400.00	
Reduce start-up time	350.00	350.00	350.00	350.00	
TOTAL BENEFITS	$1,650.00	$1,650.00	$1,650.00	$1,600.00	

Comparing the costs and benefits over four years:

	Year 1	Year 2	Year 3	Year 4	TOTAL
Costs	$2,450.00	- -	- -	- -	$2,450.00
Benefits	1,650.00	1,650.00	1,650.00	1,600.00	6,550.00
PROFIT	$ –800.00	$1,650.00	$1,650.00	$1,600.00	$4,100.00

In four years, the new equipment will pay back the original cost and generate a benefit of $4,100.00.

7. FMEA

FMEA stands for *Failure Mode and Effect Analysis*, which is a "systematic method of identifying and preventing product and process problems before they occur." The FMEA process is a way to identify failures, effects, and risks within a process or product, and then eliminate or reduce them (Exhibits C-8a, b, and c).

■ Exhibit C-8a **Sample FMEA form**

FAILURE MODE AND EFFECT ANALYSIS		
Rating	Description	Potential Failure Rate
10	Dangerously high	Failure could injure customer or employee
9	Extremely high	Failure would create noncompliance with federal regulations
8	Very high	Failure renders the unit inoperable or unfit for use
7	High	Failure causes a high degree of customer dissatisfaction
6	Moderate	Failure results in a subsystem or partial malfunction of the product
5	Low	Failure creates enough of a performance loss to cause customer to complain
4	Very low	Failure can be overcome with modification to the customer's process or product, but there is minor performance loss
3	Minor	Failure would create a minor nuisance to the customer, but the customer can overcome it in the process or product without performance loss
2	Very minor	Failure may not be readily apparent to the customer, but would have minor effects on the customer's process or product
1	None	Failure would not be noticeable to the customer and would not affect the customer's process or product

■ Exhibit C-8b **Sample FMEA form**

FAILURE MODE AND EFFECT ANALYSIS		
Rating	Description	Potential Failure Rate
10	Very high: Failure is almost inevitable	More than one occurrence per day or a probability of more than three occurrences in 10 events
9		One occurrence every three to four days or a probability of three occurrences in 10 events
8	High: Repeated failures	Repeated failures: One occurrence per week or a probability of five occurrences in 100 events
7		One occurrence every month or one occurrence in 100 events
6	Moderate: Occasional failure	One occurrence every three months or one occurrence in 1,000 events
5		One occurrence every six months or six occurrences in 10,000 events
4		One occurrence every year or six occurrences in 100,000 events
3	Low: Relatively few failures	One occurrence every one to three years or six occurrences in 10,000,000 events
2		One occurrence every three to five years or two occurrences in one billion events
1	Remote: Failure is unlikely	One occurrence in greater than five years or one occurrence in one billion events

8. FLOWCHARTING

This is a graphic representation of a process (Exhibit C-9). Visual step-by-step pictures show a series of events that produce an output. It breaks down the process into smaller workable steps. It points out missing or unnecessary steps in a process and points out areas for process improvement. Flowcharting shows how each step in a process flows into and connects with the next.

■ Exhibit C-8c Sample FMEA form

FAILURE MODE AND EFFECT ANALYSIS

Rating	Description	Definition
10	Absolute uncertainty	The product is not inspected or the defect caused by failure is not detectable
9	Very remote	Product is sampled, inspected, and released based on Acceptable Quality Level sampling plan
8	Remote	Product is accepted based on no defectives in a sample
7	Very low	Product is 100% manually inspected in the process
6	Low	Product is 100% manually inspected using go/no-go or other mistake-proofing gauges
5	Moderate	Some statistical process control (SPC) is used in process
4	Moderately high	SPC is used and there is immediate reaction to out-of-control conditions
3	High	An effective SPC program is in place with process capabilities
2	Very high	All products 100% automatically inspected
1	Almost certain	The defect is obvious or there is 100% automatic inspection with regular calibration and preventive maintenance of the inspection equipment

■ Exhibit C-9 Flowchart of a packing process

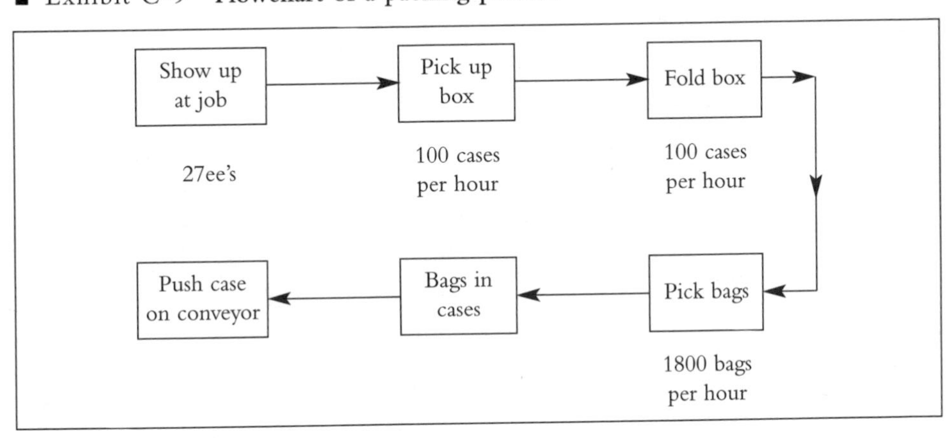

9. SCATTER DIAGRAM

This is a visual representation of what happens to one variable when another variable changes (Exhibit C-10). It analyzes the relationships between two variables and may have a cause-and-effect relationship. The scatter diagram:

- is used to test whether or not a cause-and-effect relationship exists between the two variables.
- indicates if there is a positive relationship, negative relationship, or no relationship between the two variables.

■ Exhibit C-10 **Sample scatter diagrams**

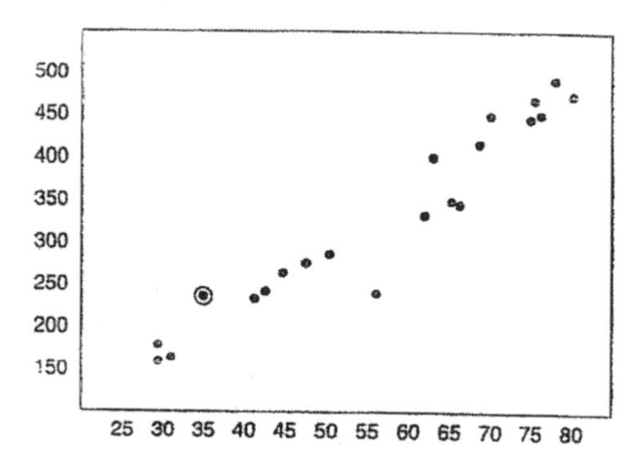

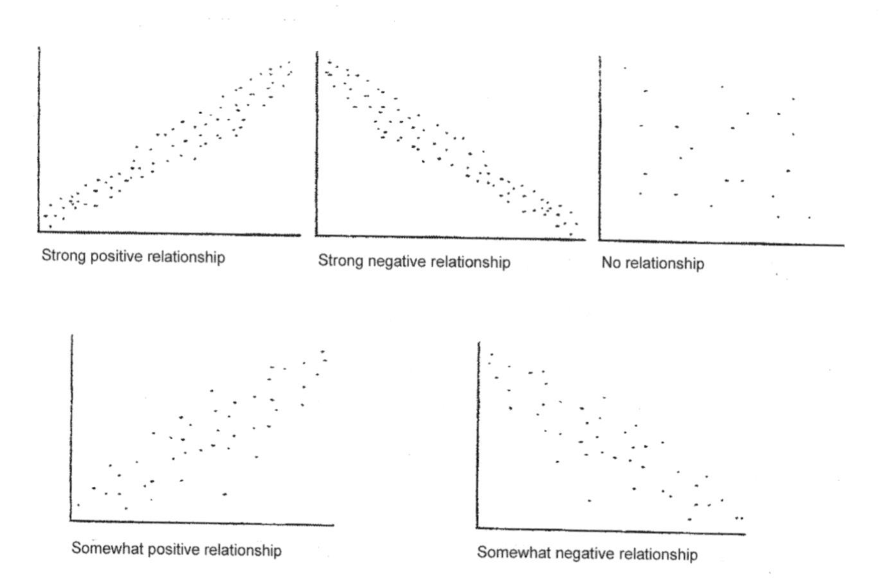

Appendix D

HOW COMPANIES MEASURE SAFETY: CASE STUDIES

Case Study 1

The company highlighted in this case study was a metal fabricator headquartered in the Midwest. It had 150 employees spread over three sites. There were three superintendents and ten supervisors. Many of the fabrications were quite heavy.

Management chose to retain the decision as to what performances they felt were essential to achieving what they wanted in safety. Executive management did not compare themselves to other companies. They believed that the following six items were things they needed supervisors to do regularly:

- inspect and improve their work area
- train their people
- investigate accidents for causes
- make one-on-one safety contact with each worker
- hold monthly safety meetings with employees
- provide safety orientation to new employees.

Management assigned point values to each of these six activities based on their perceived relative importance (see Exhibit D-1).

■ Exhibit D-1 **Measurement of activities**

Item	Points
Departmental inspections	25
Training or coaching (e.g., 5-minute safety talks)	25
Accident investigations	20
Individual contacts	20
Meetings	5
Orientation	5
Total	100

■ Exhibit D-2 **SCRAPE activity report form**

Department ______________________ Week of ____________ Points

(1) Inspection made on ________________ # corrections _________ ______

(2) 5-minute safety talk on ______________ # present ____________ ______

(3) # accidents ________________ # investigated ______________ ______

Corrections ____

____ ______

(4) Individual contacts:

Names ____

____

____

____ ______

(5) Monthly employee safety meeting held on ________________ ______

(4) New employees (names): Oriented on (dates):

__________________________ ____________________

__________________________ ____________________

__________________________ ____________________ ______

Supervisors document what was accomplished each week by completing the form shown in Exhibit D-2. From this weekly data, management generated a weekly report (Exhibit D-3) which showed the overall safety performance of each department's supervisors for that particular week.

The average of the supervisor's weekly scores were the only performance measure used in the performance appraisal system. Since the company had traditionally used financial bonuses, the average weekly supervisors' safety scores impacted the size of their annual bonus.

■ Exhibit D-3 **Example of management's report**

WEEKLY REPORT ON SUPERVISORS

Week of ______________________

	Activity						
Department	Inspections (25)	5-min. talks (25)	Accident* investigations (20)	Individuals contacted (20)	Monthly meetings (5)	Employees orientated* (5)	Total rate (100)
A	25	15	20	15	5	5	85
B	5	10	20	5	5	5	50
C	25	10	5	5	5	5	55
D	15	25	20	20	0	5	85
E	10	5	0	0	5	0	20
F	20	20	15	5	0	0	60
Average	17	14	13	8	3	3	58

*Maximum points are awarded if either there are no accidents to investigate or no new employees to orientate.

In the first year, to get attention focused on safety, a large percentage of the bonus was based on safety performance. Once habits were set, the bonus percentage for safety was reduced.

Middle management's role was to receive the supervisors' forms and follow up on them, occasionally spot-checking the quality of performance through their own inspections and talking to workers.

Results

In all management meetings, the weekly and monthly results were reviewed in detail. In the first year of the system, accidents were reduced by over 50 percent. In each of the succeeding three years, they continued to decrease at an average of over 30 percent per year.

This company manufactured store fixtures and shelving. It had no safety program in effect, and its safety record was worse than average. After several OSHA inspections, to avoid further fines, a part-time safety director was finally brought in from a local university. Working with the company training director, the company changed its direction from compliance alone to line-management accountability.

It was recently cited by the National Safety Council as one of the ten safest companies in the United States; injury reduction was across the board from power-press injuries to back strains.

Case Study 2

The focus of this case study was a division of the Burlington Northern Railroad. The study was part of a national study on safety in the industry. The measures here were to test what observations, contacts, and positive reinforcement could do. The company had 250 employees with fourteen first-level supervisors and two division managers.

Management wanted a simple system aimed at accidents that were caused by human behavior, such as unsafe acts. After reading the current leadership and management books, they decided to try to change worker behavior by using positive behavior reinforcement instead of the more traditional approach involving enforcement of rigid rules.

All supervisors were required to attend a six-hour training course, which was developed and presented by the company's managers. This course consisted of the subjects shown in Exhibit D-4.

The supervisors were required to actually *use* the training they received when they returned to the job. In each case, the division manager was asked to devise a simple system of accountability where each supervisor had to report weekly on how they had used the concepts that were discussed during training. In each case, the supervisors were asked to observe workers as usual and to report the number of observations they made each week to their managers.

When the supervisors made an observation, they were required to follow certain procedures. If the worker observed was working unsafely, the supervisor was asked to deal with the infraction in the normal manner. If, however, the worker was working safely, the supervisor was required to contact the worker and positively reinforce that desirable safe behavior. It was believed that this type of positive behavior reinforcement would make the worker more likely to continue to work safely in the future.

Each week supervisors reported on a 3 x 5 note card the number of total observations made and the number of positive reinforcements given. No names were asked for or recorded, merely the number of times that observations were made and the total number of positive reinforcements that were given.

■ Exhibit D-4 **Supervisory training-course subjects**

HOUR 1–2	INTRODUCTION
	Overview
	Introduction to the session
	Pretests and questions
	
	WHY TRY A DIFFERENT APPROACH TO SAFETY
	Updating safety theory
	• What we've always done vs. what we now know
	• What an accident is, and what causes them
	    ◦ Unsafe acts/conditions vs. human error
	• What we can do to prevent accidents
	    ◦ The 3 Es of safety vs. the 3 behavior changers
	• What other companies and industries have found

HOUR 3	WORKER MOTIVATION AND WHAT WE KNOW ABOUT IT
	What motivation is
	• A person doing whatever is necessary to satisfy current needs
	What needs are
	• The major theories and concepts
	    ◦ Needs of most workers
	    ◦ Need change
	    ◦ Motivators and dissatisfiers
	• What the research shows

HOUR 4	INFLUENCING WORKER BEHAVIOR
	Your three choices
	• Motivation – the environment
	• Attitude change
	• Changing behavior
	Motivation—the easiest way
	• What influences workers usually
	    ◦ The peer group
	    ◦ You and your style of leading
	    ◦ Your credibility (particularly in safety)
	    ◦ Your attitude toward safety (what they see)
	    ◦ How you measure and judge them
	    ◦ What they think your priorities are
	    ◦ The organizational climate

HOUR 5	WHAT YOU CAN DO Understand each worker Understand the group of workers you supervise Use the motivators
HOUR 6	USING POSITIVE RECOGNITION

The supervisors' weekly observation numbers were reported directly to the division manager. The division manager's role was to make sure that each supervisor sent in a 3 x 5 card weekly with the observation numbers on it and to spot-check the quality of the positive reinforcements given by the supervisor through inspecting the workplace and talking to workers. The supervisor's reward was the feedback received from the division manager on the quantity and quality of positive behavior reinforcements given out by the supervisor.

Results

The result of this rather simple program was a 40 percent reduction in unsafe behavior in the first three months, which continued throughout the life of the program. Also, there was a 51 percent reduction in accidents the first year. This positive reinforcement program is still in place at this company today, with accidents declining continuously each year.

These results track with similar programs at three other railroads in the same time period: The Southern Railway System; The Illinois Central; and The Duluth, Missabe, and Iron Range Railway Company. In all four, the only measurement that was seriously considered was safety sampling by the four company safety professionals.

Case Study 3

The company studied here was a contractor with 50 employees and three supervisors who reported directly to the owner. There was no structured safety program, but the owner wanted to improve his company's safety record.

The owner felt that the key to safety performance was to get each supervisor to accept ownership of the safety program. They decided to use an SBO accountability system. In an SBO system, the role of the company's owner is to ensure supervisory success in meeting the established objectives by removing barriers and providing supervisors with the necessary training.

Each supervisor formed an agreement with the company's owner about the safety activities for which he or she would be responsible. These activity objectives were put in writing. Exhibit D-5 shows the safety activities that one of the supervisors was responsible for.

Results

This company achieved a 62 percent reduction in injuries in the first year, 40 percent in the second year, and 35 percent in the third year.

■ Exhibit D-5 **Safety activity objectives**

SAFETY ACCOUNTABILITIES

Supervisor: ______________________________

Manager: ______________________________

1. Inspect my department for physical hazards using the checklist once each week. Checklist to be filed in the department.
2. Contact each employee individually at least once each week and discuss something about safety relevant to their job.
3. Provide at least one positive comment to each employee following an observation of their performance—one per employee per week.
4. Inspect for OSHA violations once per week using the OSHA checklist. Checklist to be filed in the department.
5. Attend the manager's safety meeting each month.
6. Document the above with a weekly report submitted to my manager.

Case Study 4

A large Texas facility of the DuPont Corporation had 1,800 employees at one location. Management believed that there were some safety issues that had not been addressed. Half of the organization is managed using the total quality management (TQM) concept. In these areas, there are no supervisors, and a team runs the department. The other half of the company uses a more traditional style of management, which employs supervisors. There are a total of fifty supervisors who work in this half of the operation.

The company had a good safety record. However, the plant manager felt there was still room for improvement. The company identified twelve activities that seemed to show weaknesses. If supervisors

■ Exhibit D-6 **SBO accountability system**

SAFETY OBJECTIVES

1. Reinforcing good safety effort (providing positive behavior reinforcement for doing the job safely).
2. Promoting the philosophy that shortcuts are not necessary to maintain production and will not be permitted.
3. Visibly promoting the philosophy that safety is important and *equal* to all other considerations (cost, quality, production, human relations).
4. Spending more time in the field to gain first-hand knowledge of safety problems and attitudes.
5. Insisting on prompt action from technical, mechanical, or production to correct *known* safety problems (quit living with known safety problems).
6. Promoting good housekeeping.
7. Having better personal knowledge of people with poor safety attitudes and working with immediate supervision to correct.
8. Promoting familiarity with safety procedures pertinent to our work.
9. Promoting the wearing of required safety equipment and following safety and operating procedures at all times.
10. Visibly demonstrating to subordinates your standards for safety performance and your acceptance of your responsibility for their safety.
11. Promoting more attendance of area safety meeting by higher supervision.
12. Promoting preparation of higher quality safety meetings.

and teams would concentrate in these areas, it was felt that the safety record could be further improved.

Using the SBO accountability system, each supervisor (or team) was required to select three of the twelve safety critical areas (Exhibit D-6) and set specific measurable objectives that they could work on to improve in that area. The specific activities were left up to the supervisors and teams to decide. After that, they needed to reach an agreement on these objectives with their managers.

The role of the middle manager was to ensure the success of the supervisors and teams by providing them with the necessary support, training, equipment, and funding to accomplish their objectives.

Results

Management tracked progress in the twelve areas through sampling of behavior. They also changed their incident investigation procedure to specifically ask about the twelve areas. With a good record to start with overall, incidents declined over 20 percent in the first year.

Since the company had always had a superb safety record, the management was pleased with the overall results, and pleased with the reductions of incidents in the defined areas. In this case, no attempt was made to correlate the data beyond what they were looking for.

Case Study 5

The company at the center of this case study was a Frito-Lay Corporation facility with 500 employees and three levels of management. Their first-line supervisors typically supervise twenty to twenty-five employees. The next level of management is the department managers. At the top is the plant manager.

The company decided that the MENU accountability system fit their management style the best. MENU retains some mandatory activities for supervisors but does allow for some flexibility.

Exhibit D-7 shows the required safety tasks that all first-line supervisors are required to perform. In addition to their required tasks, a MENU of optional activities is provided to each supervisor. From the

■ Exhibit D-7 **First-line supervisor accountabilities**

GENERAL

The key accountability of the first-level manager is to carry out the tasks defined below.

TASKS

Required tasks are:

- Hold a monthly safety meeting with all employees
- Include safety status in all work-group meetings
- Inspect department weekly and write safety work orders as required
- Have at least five one-to-one contacts regarding safety with employees each week
- Investigate injuries and accidents in accordance with managing safety guidelines within 24 hours

In addition, in agreement with department head:

- Select at least two other tasks from a provided list and agree on what measurable performance is acceptable
- Report on these activities weekly

WEEKLY SAFETY REPORT

The first-level manager shall prepare and distribute a weekly safety report in accordance with the format shown.

MEASURE OF PERFORMANCE

- Successful completion of tasks

REWARD FOR PERFORMANCE

- Safety will be listed as one of the key measures on the accountability appraisal form

MENU, the supervisor is required to select two additional tasks for which he or she will be responsible. A supervisor's performance was then simply determined by whether or not any tasks were completed.

Exhibit D-8 shows the required tasks for department managers. All these tasks are required, with the exception of the third task, which asks the department manager to select an activity to perform that would demonstrate his or her commitment to safety. The department manager would work with the plant manager to reach a mutual agreement on the specific number of inspections and one-on-one safety contacts that would be made each month. The department manager's performance measurement was a combination of completing assigned tasks, the results of a departmental safety audit, and the accident record realized by the manager's department over the preceding year.

■ Exhibit D-8 **Department manager accountabilities**

GENERAL

The key accountability of the department manager is to ensure that the plans and programs of the company safety system are carried out in their area.

TASKS

- Review reports from their area on task accomplishments and act accordingly
- Assess task performance defined for subordinate managers and feedback as appropriate
- Engage in some self-defined tasks that can readily be seen by the workforce as demonstrating a high priority to employee safety
- Develop safety management knowledge and skills in subordinate managers
- Make one-to-one safety contacts with hourly employees
- Participate in department safety inspections

MONTHLY SAFETY REPORT

Department managers shall prepare and distribute a monthly safety report in accordance with the format shown.

MEASURES OF PERFORMANCE

- Successful completion of tasks
- Safety audit results for area(s) of accountability
- The 13-month rolling total injury frequency record for area(s) of responsibility

REWARD FOR PERFORMANCE

- Safety will be listed as one of the key measures on the accountability appraisal form

■ Exhibit D-9 **Plant manager accountabilities**

GENERAL

The key accountability of the plant manager is to ensure plant safety performance.

TASKS

- Hold a monthly plant safety meeting for all managers
- Make a plant safety inspection and observation tour once each month
- Review reports of subordinate and supervisory performance and act appropriately
- Review monthly safety report and act accordingly
- Develop safety management knowledge and skills in subordinate managers

MEASURES OF PERFORMANCE

- The plant's overall safety audit score
- The total cost of workers' compensation claims
- The plant's 13-month rolling injury frequency record

REWARD FOR PERFORMANCE

- Safety will be listed as one of the key measures on the accountability appraisal form

Exhibit D-9 shows the plant manager's responsibilities. There are three measures of the plant manager's performance. First, the company's safety department conducts an annual audit of the safety program and does a safety inspection of the workplace. The results of this audit are scored, and the plant manager is responsible for the overall score. Next, the plant manager is responsible for the total cost of workers' compensation claims that occurred during the year. Finally, the plant manager is held accountable for the average injury frequency rate the plant had during the preceding thirteen months.

Results

At the outset, the organization's safety record placed it in the upper third of companies in the food industry with the highest accidents and injuries. A goal was set for the plant to achieve a 40 percent reduction in total recordables per year. The first year a 36 percent reduction was attained, the second year showed a 50 percent reduction, and a 41 percent reduction was attained in the third year. At the end of the three-year period, the plant had the second best record in the food industry. All types of injuries were reduced across the board.

Case Study 6

The final case study concerns perception surveys at Minnesota Power. A division of ALLETE, Minnesota Power provides electricity in a 26,000-square-mile electric-service territory located in northeastern Minnesota; it supplies retail electric service to 135,000 retail customers and wholesale electric service to sixteen municipalities. Minnesota Power has used perception surveys for the last four years to track progress in their safety efforts. As discussed earlier, a perception survey serves to establish a baseline and diagnose what needs to be fixed. Some organizations do a survey annually to assess progress, then they fix the problems diagnosed.

Minnesota Power has an 85% positive employee-only score overall and has only one score below the 70% level. Each year the scores have improved; over five years some categories have improved 15%, and all twenty categories show improvement for the period. In addition, the difference between levels (employee to supervisor) has dropped from a 10% difference in 2000 to a 2% difference in 2004.

Below is a 2004 comparison of Minnesota Power (employee-only positive responses) to the national averages.

Category	National	Minnesota Power
Recognition	61.9%	72%
Discipline	61.5%	66%
Inspection	62.3%	83%
Supervisor training	71.4%	84%
Alcohol/drugs	72.5%	78%
Employee training	77.0%	87%
Quality supervision	76.6%	88%
Involvement	74.6%	81%
Operation process	72.5%	78%
Attitude	76.6%	83%
Support	74.6%	87%
Management credibility	77.2%	89%
Goals	78.0%	86%
Climate	76.6%	87%
Safety contacts	82.0%	92%
Hazard correction	78.1%	91%
New employee orientation	79.8%	89%
Awareness	72.4%	80%
Communication	80.5%	87%
Accident investigation	85.8%	95%
Overall	74.2%	84%

Minnesota Power is one of many organizations that have used the perception survey. They are now using improvement in perception survey results as a primary measure of safety performance, slowly moving away from injury measures. After five years their performance has improved to a point where the injury count does not seem to be as good an indicator of system effectiveness as opinion survey results. In addition, the surveys help to direct focused improvement each year.

Index